DEVELOPING ECOLOGICAL CONSCIOUSSNESS

DEVELOPING ECOLOGICAL CONSCIOUSNESS

The End of Separation

Second Edition

Christopher Uhl

ROWMAN & LITTLEFIELD PUBLISHERS, INC.
Lanham • Boulder • New York • Toronto • Plymouth, UK

Published by Rowman & Littlefield Publishers, Inc.
A wholly owned subsidary of The Rowman & Littlefield Publishing Group, Inc.
4501 Forbes Boulevard, Suite 200, Lanham, Maryland 20706
www.rowman.com

10 Thornbury Road, Plymouth PL6 7PP, United Kingdom

Copyright © 2013 by Rowman & Littlefield Publishers, Inc.

All rights reserved. No part of this book may be reproduced in any form or by any electronic or mechanical means, including information storage and retrieval systems, without written permission from the publisher, except by a reviewer who may quote passages in a review.

British Library Cataloguing in Publication Information Available

Library of Congress Cataloging-in-Publication Data
Uhl, Christopher, 1949–
    Developing ecological consciousness : the end of separation / Christopher Uhl. — Second edition.
    pages cm
    Includes bibliographical references and index.
    ISBN 978-1-4422-1831-4 (cloth : alk. paper) — ISBN 978-1-4422-1832-1 (pbk. : alk. paper) — ISBN 978-1-4422-1833-8 (electronic) 1. Ecology. 2. Sustainable development. 3. Restoration ecology. 4. Ecological assessment (Biology) I. Title.
    QH541.U44 2013
    577—dc23

                                                                2012050095

♾™ The paper used in this publication meets the minimum requirements of American National Standard for Information Sciences—Permanence of Paper for Printed Library Materials, ANSI/NISO Z39.48-1992.

Printed in the United States of America

For young people everywhere
Including my amazing students at Penn State
And my beloved Jake and Genny and Katie

CONTENTS

Preface to New Edition   ix

Acknowledgments   xvii

PART I: EARTH, OUR HOME   1

1 Discovery: Belonging to Something Greater Than Ourselves   5

2 Coming to Awareness: A Living Planet   29

3 Cultivating Community: Intimacy with Earth's Web of Life   55

PART II: ASSESSING THE HEALTH OF EARTH   85

4 Listening: Gauging the Health of Earth   89

5 Courage: Facing Up to the Unraveling of the Biosphere   117

6 Living the Questions: Discovering the Causes of Earth Breakdown   147

PART III: HEALING OURSELVES, HEALING EARTH   173

7 The Old Story: *Economism* and Separation   177

8 Birthing a New Story: The Great Turning   205

Epilogue   237

Notes   243

Index   261

About the Author   269

PREFACE TO NEW EDITION

It's not enough to invent new machines, new regulations, new institutions. We must develop a new understanding of the true purpose of our existence on earth. Only by making such a fundamental shift will we be able to create new models of behavior and a new set of values for the planet.

—Vaclav Havel, President of Czech Republic[1]

I wrote the first edition of this book twelve years ago when I was in the Pacific Northwest on a sabbatical leave from Penn State. I chose that region because it appeared to be popping with fresh ideas and initiatives, both ecological and social. Between bouts of writing, I went out to explore what was going on. At the time I was like a sponge, ready to soak up anything and everything that was new and different. I was open to most everything—workshops on nonviolent communication, symposia on climate change, retreats on insight meditation, classes on social networking, drumming circles, field trips on insect ecology, seminars on environmental advocacy, and much more. I participated fully, took copious notes and, in a state of unbridled enthusiasm, incorporated much of what I learned and experienced into the first edition of this book. I was so excited by what I was learning that I didn't want to leave anything out. The result was a variegated and, I hope, inspiring mix of ideas, concepts, stories, revelations, perspectives, questions, and practices.

MY RATIONALE FOR CREATING A NEW EDITION

In 2007, when my editor invited me to create a new edition, I declined, believing that the book was fine just as it was. Five more years went by and in 2012 my editor, once again, suggested a rewrite. This time I consented, imagining that it would involve a bit of updating, some cutting and pasting, and perhaps the inclusion of some new material. I anticipated finishing the job in a month or two. But when I sat down to carefully read my book for the first time in more than a decade, I quickly realized that my understanding of environmental issues had deepened considerably since writing the first edition, especially as regards the seriousness of climate change, the deteriorating condition of Earth's oceans, the shortcomings of our economic paradigm, the psychology of consumption, and the fundamental relational nature of life. Along with this, I had gained a much fuller understanding of ecological consciousness and how it can be engendered.

Somewhat reluctantly, at first, I conceded that much more than a little freshening up would be required for a new edition. Indeed, the time had come to significantly rethink; renew; reorganize; and, in fact, reimagine this book.

If you know this work from its first edition, you will see immediately that much has changed, though the basic structure of the book—replete with foundations, stand-alone boxes, reflections, questions, and practices—remains intact.

Yes, there has been some serious culling—for example, nixing much outdated and extraneous material—as well as significant reorganizing, such as adding, removing, and rearranging sections; and all of this has been done with an eye to creating a more coherent and compelling narrative.

Even with the addition of new material, I have reduced the overall length of the book but, I trust, not in ways that earlier readers will lament. Indeed, a dozen years of working with the first edition in the classroom has afforded me ample time to test out content, tone, and message and to cull material that had limited appeal or that was sometimes more diversionary than essential.

LEAVING *SUSTAINABILITY* BEHIND: THE NEED FOR A NEW STORY

Previous readers will note that the concept of sustainability that figured prominently in the original book is all but banished from this edition. In the first edition, my attachment to *sustainability* was so strong that I included this word in the book's subtitle. But, in the interim, I have come to see that sustainability, insofar as it seeks to figure out ways to sustain our current way of life, is profoundly problematic.

As I now see it, the concept of sustainability is mired in the consciousness of *if only*. For example, *if only* we all recycle our empty beverage containers and compost our kitchen scraps, everything will be OK; *if only* we all drive a Prius everything will be OK; *if only* we fly a little less . . . *if only* we choose cloth diapers instead of paper . . . *if only* we heat our homes with geothermal technologies . . . *if only* we buy *energy-star* electronics. . . . You get the idea. But bundle up all these *if only* contingencies and put them into practice and their net impact would be small relative to the enormous momentum and associated ecological impacts of today's careening consumer economy.

All the *dos and don'ts* of the sustainability movement will not save us. The problem, I now see, is that sustainability, for all its good intentions, is part of our old story—the story that tells us that we can keep growing (as long as it's smart growth) and keep consuming (as long as we recycle) forever. It's a story that tells us that technology—so long as it's green—will solve our problems; and that happiness resides in *having* belongings, not in *the experience of* belonging. This story is no longer tenable. The clamor around sustainability is a last-ditch attempt to have our cake and eat it too. Who wouldn't want that?

Something bigger and deeper and more soulful than the shortsighted notion of sustainability is needed if humankind is to flourish in partnership with Earth. It will not be enough to prop up our old "business as usual" story with good intentions and green trimmings. That story is crumbling, as manifested in the widespread deterioration of Earth's ecosystems, climate, and biota.

This book is an invitation to see life, both personal and societal, through the lens of a new story. We each live many-storied lives. When we speak, we are telling stories; when we listen, we are creating stories about both what we hear and about the one who is speaking.

There is now evidence from many quarters that humankind is in the throes of birthing a new story that is infinitely more joyful, life-affirming, and spirit-expanding than our old story. This new story challenges us to wake up and, in so doing, to see ourselves, not as *apart from* Earth (as we now tend to do), but instead, as *a part of* Earth—literally a *part of* the body of Earth!

MY OWN AWAKENING

Three decades ago, when I began teaching environmental science at Penn State, I saw wounds everywhere—clear cuts, acid rain, ozone thinning, polluted rivers, toxins in our food, horrendous waste, wrenching wars—a world, seemingly, hurtling toward its own demise. In recounting these wounds to my students I told my *story* of environmental science. At the time, I was so invested in my story that it barely occurred to me that I could, if I chose, center my course around a different story.

The more I inhabited my doom-and-gloom story the more I became sad, angry, and indignant. Indeed, back then, a good class for me was one where I delivered a rant about the latest environmental calamity. I say "good class" because my venting enabled me to experience some measure of personal catharsis. But the larger truth, now self-evident, is that by passing my angst on to my students, I was acting in an insensitive and extremely self-absorbed way.

Can you imagine being me—Dr. Death—year after year tracking the deterioration of Earth's vital signs? Or perhaps worse, can you imagine being a young person—filled with energy, fire, and goodwill—having to sit in a room with four hundred of your contemporaries, receiving information about how the planet is screwed and there is probably nothing that you can do about it?

This phase of my teaching ended when I realized, to my horror, that rather than engendering an ethic of stewardship and caring for Earth, my negativity was, more likely, doing just the opposite. It took me a long time to grasp this. One important epiphany occurred on the last day of class at the end of my sixth year of teaching my environmental science course. From my perspective it had been a good year. I had incorporated some new material into the course and I was becoming a better lecturer. As I stood in a self-congratulatory stance, my students sat with their heads bowed, laboring over their final exam. As the hour wound down, they came up, one by one, to hand me their test sheet. I was feeling lighthearted and ready to wish them well and to thank them for taking my course. My students, on the other hand, appeared sullen and downtrodden. Only a handful even made eye contact with me. Here I thought I was at the peak of my game, but something was seriously amiss.

Sometime later, feeling forlorn, I grabbed my backpack and headed to the mountains for a weeklong *walkabout*. The simple act of sauntering through Penn's Woods reminded me of the wonder and peace—the sense of kinship and full-body delight—I had known on sojourns in the wild as a young man. And it was during this retreat to the woods that I realized that I was teaching environmental science upside down. I was asking my students to care about something—Earth—with which most of them had little contact—little connection—almost no felt relationship. Like the rest of us, my students had become indoor people, domesticated, out of touch with rock and soil, free-running water, unfiltered sunlight, blowing wind, and wild creatures. How absurd of me, then, to expect them to care about an Earth with which most had very little relationship!

Further, I realized that by grounding my course in an ethos of fear and guilt and doom, I was engendering hopelessness, revulsion, even numbness, in my students. Only then did it occur to me to consider how it would be if I were to ground my teaching in awe, love, compassion, wonder, and possibility. Yes, what if I sought to tell a story that could lead my students to fall in love with Earth? It was this question, in concert with my angst, that allowed me to birth a new story as told in this book.

THE STORY I TELL IN THIS BOOK

This book and my work as a teacher of environmental science has been motivated by the simple question: How might we make sense of these times into which we have been born and, then, use this understanding to create lives filled with meaning and purpose? In the eight chapters that follow, I offer my best answer to this question.

I begin, in chapter 1, with the perennial question: How is it that human beings even exist? I explore this question by laying out the cosmological story of where all the things of creation—the stars and planets, the bacteria and fungi, the plants and animals (including us)—have come from. I tell this story through the lens of science, mindful, all the while, that science is just one way of knowing, or as Einstein said, "Science without religion is lame; religion without science is blind."

In chapter 2 I introduce the metaphor of *Spaceship Earth*, first coined by the American visionary Buckminster Fuller. It's an accurate moniker; Earth *is* our *spaceship* insofar as it contains, within its boundaries, all that we require to live; absent Earth's gifts, we cannot survive. We all know the rules for driving, but what about the rules for living successfully on our home planet? Do you know enough to earn your "learner's permit?" You'll find out in chapter 2.

In chapter 3 I will invite you to consider that the essence of life is relationship. Indeed, you are alive right now by virtue of billions of relationships being enacted in your body second by second. When you complete this chapter it is my hope that the literal truth of the statement, *Earth is your larger body*, will be fully apparent to you.

These first three chapters comprise part I. I think of them as compass coordinates helping us to see Earth—the planet that has birthed us—through the eyes of wonder, gratitude, awe. Next, in part II (chapters 4–6) I present evidence revealing that Earth, our larger body, is ailing.

Chapter 4 examines the beleaguered condition of Earth's oceans and the tattered state of Earth's terrestrial environments, while chapter 5 focuses on the disheveled state of Earth's atmosphere and the proliferation of toxic chemicals in Earth's soil, water, and air, not to mention in the very tissues of our bodies. The evidence is incontrovertible: As Earth sickens, so do we.

I conclude part II, in chapter 6, exploring the root causes of the ecological crisis now plaguing life on Earth. In its simplest formulation, the crisis is the result of humans taking more from Earth than our planet has to offer. Indeed, all our ecological problems—from dwindling freshwater supplies through rising amounts of greenhouse gases in the atmosphere to the loss of forests and the depletion of topsoil—trace back to our failure to respect Earth's limits. One outcome of our intransigence may well be that Earth will simply slough us off, just as a dog might shake off annoying fleas. After all, Earth really doesn't need us; we, on the other hand, desperately need Earth for our continued survival.

So what are we to do? Is there a way out? That's what part III (the final two chapters) is about. In chapter 7, I posit that it is both our story about Earth (i.e., that Earth is simply a storehouse of resources put here for our relentless exploitation) and our story about our purpose (i.e., that a successful life results from working hard to make money to buy things that will bring happiness) that is leading to the sickening of Earth and the diminishment of our spirits. Rather than this story, the time has come for a new story that reflects and celebrates the fundamental relatedness of life, a story that acknowledges the basic goodness of the human and that calls on us to discover that what we do and give to each other, we, ultimately, do and give to ourselves.

In this book's final chapter, I describe how this new story of kinship and community and deeper purpose is already emerging. Indeed, amid destruction, a new world is being birthed—a world where our fundamental needs for food and energy and shelter will be met, to a significant degree, in the bioregions where we dwell; where we will each give our gifts freely in the service of the common good; and where we will take a respectful, even reverential, stance toward Earth as our larger body.

In sum, each of this book's three parts contains a necessary ingredient for a good story. In part I, I introduce the principal characters—namely, the generative universe that has birthed us, the Earth and the Sun that nurture us, and the web of life that we are enmeshed in. In part II, I describe how we, an upstart species, are presently wreaking havoc on Earth. Then, in part III, I suggest a way that this drama might be resolved that is contingent on our ability, as present-day humans, to transform our consciousness. Will there be a part IV someday? I suppose that depends on us. . . .

PRACTICES TO AWAKEN BOTH OUR INTELLECTS AND OUR HEARTS

It would be possible to read this book, nodding and grunting and grumbling along the way, and then to put it down and move on to one's next agenda item. I know. I have read many books, and though they may have touched me momentarily, I moved on, often unmarked in any enduring way. As a foil to this tendency, I offer "applications and practices" at the end of each chapter. The idea is to provide suggestions for what one might actually do to experience—in an embodied way—the core ideas of each chapter. Implicit in these practices is the recognition that we humans are much more than mere brains on the end of a stick. We are enfleshed beings—with the capacity for love, forgiveness, compassion, grief, hope, wonder, intuition, despair, insight, joy, celebration—searching for meaning in a universe imbued with mystery.

Writing this second edition, I have felt at times akin to a mother who gives birth to one child and, then, slowly over time, becomes ready to birth a second child. Now, after almost nine months, some heavy and frustrating, others gratifying, I am done. Yes, I am ready to let go, to release, to shout *hallelujah*. I have done my best to write in a manner that is honest. I apologize for my shortcomings. This story, after all, just like life itself, is a work in progress.

ACKNOWLEDGMENTS

I offer my thanks to Na Eun Yoon and Brad Podolski for evaluating an early draft of this book; to Melissa DiJulio for reviewing and proofing selected chapters of the book; and to the students in my 2012 BiSci 03 Honors course (where this book was piloted) for their thoughtful comments and suggestions.

My thanks also go to Gabrielle Bedeian, for helping me decide what to retain and cut from the first edition, as well as her advice on how to make this edition more engaging.

My sincere appreciation also goes to Gregory Lankenau, who did a comprehensive critique of the book, making hundreds of suggestions regarding phrasing, tone, audience, and organization, as well as gently cajoling me to weave the book's central themes into a coherent whole.

I am grateful, as well, to my colleagues Robert Burkholder (Penn State) and Coleen O'Connell (Lesley University) for reading the text and offering both insightful comments and encouragement. As well, my thanks go to Charles Fisher and Eric Post for reviewing the material on oceans (Fisher) and climate change (Post).

In addition, I am grateful to John Oat for creating four new figures for this edition; and to Jean Forsberg for granting permission to grace this book's cover with one of her extraordinary paintings.

My thanks also go to Sarah Stanton, Kathryn Knigge, and Janice Braunstein at Rowman & Littlefield for their skillful means in moving this new edition through the publication process.

Finally, I offer a deep bow to my partner, Dana Stuchul, who extended moral support, good cheer, and patience during the highs and lows of this project.

Part I

EARTH, OUR HOME

The real *voyage of discovery* consists *not in seeking new* landscapes *but in having new eyes*.

—Marcel Proust

The goal of part I is to connect—specifically, to forge a connection to the cosmos, to Planet Earth, to the life around us, and to ourselves. By connecting, we can open ourselves to potentially life-altering experiences such as those described by some of the early astronauts. For example, when astronaut Rusty Schweickart was released from his space capsule on an *umbilical cord* during an early Apollo mission, he looked back to Earth, "a shining green gem against a totally black backdrop," and he realized that all that he loved was on that gem: his family; the land and rivers of his home place; art, history, culture. This jet-fighter pilot was so overcome with emotion that he wanted to "hug and kiss that gem like a mother does her firstborn child."[1]

Schweickart had another breakthrough: As he observed planet Earth from outer space, he saw that clouds did not stop at national borders to check for political ideology; and he saw that ocean currents, rivers, and mountain chains took no heed of nation-states. This red-white-and-blue American came to understand in a profound way that national boundaries are political constructs devoid of biological meaning. The natural world, built on the principle of interdependence, ignores these artificial boundaries.

This call to connect to the real world of air, soil, water, and wild beings is bewildering to many people young and old. You, yourself, might be thinking: "I already am connected! I know what I am doing!" If so, here are some questions for you: What is the foundation of our society? What makes the world go around? What keeps you alive? Is it money? Technology? Business? Government? God? Really, what is it that supports the whole show we call "society?" In pondering this question, it is common to overlook our most fundamental support systems: the Sun—that great radiance that lights up the world; water—that miracle liquid that enables life; soil—that rich substrate that is the growth medium, directly or indirectly, for all that we eat; the atmosphere—that all-permeating elixir that we draw our breath from.

This forgetting is really not so surprising. As we humans have become more specialized and dependent on technology, our separation from planet Earth has increased. This separation is literal. Here in the United States, we spend more and more time indoors, living inside boxes. Indeed, every day, almost all of us get up in the morning inside a box (house or apartment). Then, we hustle off to work or school in a small horizontally moving box (cars, buses, trains); then, we proceed to spend our days in other sorts of boxes (office buildings, schools, businesses, factories, stores). In the evening, we play it backward, moving from box to box, on our way home. Amazingly, we may pass an entire day without any direct contact with the living Earth: Never feeling unfiltered sunlight on our skin or a cool breeze ruffling our hair; never hearing the sounds of free-running water or walking on unpaved, bare earth; never once plucking a ripe berry from a bush or delighting in the flight-dance of starlings at dusk.[2]

So what? Here's what: Living our lives in boxes, as many of us now do, disconnects us from the living Earth. We become indoor people with indoor concerns, connected to our televisions and computers, substituting virtual reality for the real thing. In a word, we become *domesticated*, our wild selves increasingly ransomed to a simulacra existence.

This need not be. Part I of this book is an invitation to see with new eyes—the eyes of relationship—and in so doing to assume our birthright as vital and awakened members of the Earth community.

1

DISCOVERY

Belonging to Something Greater Than Ourselves

Indigenous peoples . . . live in a universe, in a cosmological order, whereas we, the people of the industrial world, no longer live in a universe. We in North America live in a political world, a nation, a business world, an economic order, a cultural tradition, a Disney dreamland.

—Thomas Berry[1]

When I first read these lines of Thomas Berry's, I was baffled. Could it be that "we, the people of the industrial world, no longer live in a universe"? Of course we do, but if we have no day-to-day awareness of this fact, can we truthfully claim to live in a universe? In the end, doesn't awareness count for something? Everything?

Some years ago, having spent far too much time "in a political world . . . a business world . . ." I retreated to the forested mountains of northern Pennsylvania. I established a camp, tended a fire, and roamed by day; at night I slung a hammock between two pines and listened to owls hoot and coyotes yelp before falling into fitful slumber. On one evening, I awoke because (in the spirit of the outdoors) nature called. Rising from my hammock I stumbled off into the darkness. But when I finished and turned around I couldn't see my hammock. Disoriented, I wandered about—hands out in front—feeling for hammock strings, but to no avail. I began to feel distressed, but then I took a deep breath and considered that the worst that could happen was that I would spend a few hours on a star-strewn night free to contemplate the heavens. I looked up and was reassured to see the Big Dipper showing clear in an opening of the forest canopy.

Then, I reached down and picked up two pine needles and held them in a crossed position, thereby defining a point within the Big Dipper's bowl. With this gesture, I was miming what the Hubble Space Telescope had done. Specifically, Hubble had focused its camera on a sky speck equivalent to that defined by the intersection of two pine needles held at arms' length, all the while soaking up the faint light from distant galaxies. The result was an extraordinary photograph showing nearly two thousand galaxies in one solitary speck of sky.

It would take roughly twenty-five thousand Hubble photographs to survey just the bowl of the Big Dipper. Such a survey would reveal an estimated forty million galaxies in the Dipper's bowl alone; a survey of the entire sky, in this manner, would turn up more than one hundred billion galaxies—each with billions of stars, many of those with families of planets. In this vein, Boston College professor Chet Raymo calls on all of us to:

> Go outside and hold those crossed [needles] against the night sky. Let your imagination drift away from the Earth into those yawning depths where galaxies whirl like snowflakes in a storm. From somewhere out there among the myriad galaxies, imagine looking back to the one dancing flake that is . . . our Milky Way. Galaxies as numerous as snowflakes in a storm.[2]

Following Raymo's call, I was, for a brief time that evening, living in a universe.

The theme of this chapter is "discovery." Today, by joining discoveries in the sciences with humanity's great wisdom traditions, we are able, as never before, to fathom the workings of the Earth and Sun that sustain us as well as the origins of the universe that is our larger home. In so doing, we discover that we are a part of something far greater than ourselves! Yes, whether we know it or not, we are all part of the community of life on Earth, members of the Milky Way galaxy, and participants in an expanding and complexifying universe.

FOUNDATION 1.1: THE ORIGINS AND WORKINGS OF THE UNIVERSE

Human beings are curious by nature. At an early age, we begin to ask big questions: How did I come to be? Where did planet Earth come from? What is a galaxy? Yes, it is in our nature as humans to ask probing questions.

Yet, I confess, with a mixture of regret and embarrassment, that I taught environmental science for ten years before it even occurred to me to invite students to ponder how our topic of study—planet Earth, and the cosmos within which Earth is embedded—came into existence in the first place.

The Old Story

Where did it all come from? As a way of gaining some perspective on this question, put yourself back 20,000 years to the time when our hunter-gatherer ancestors inhabited Earth. If you had lived then, what story would you have created to explain all those points of light (stars) in the night sky? Perhaps those flickering specks were the campfires of your ancestors—souls who had died and passed into the spirit world? And how far away were those "campfires"? A mile? A hundred miles? A thousand? Even today, if no one had told you, could you figure it out?

The Greek astronomers, puzzling over this, observed the Sun circling the Earth each day; and at night, they saw the moon and stars move across the sky. Based on their Earth-centric perspective, their conclusion, quite sensibly, was that Earth must be nested at the very center of a series of concentric celestial spheres. The Moon, they thought, was embedded in one sphere, the Sun in another; Mars in another and so forth, all the way to the final sphere that contained all the stars of the night sky.

Those early astronomers noted that the myriad stars embedded in that hypothetical star sphere differed in their relative brightness, and they assumed that this meant that the stars were of different sizes—bright ones

---

SUNSET OR EARTH ROLL?

It is hardly surprising that our ancestors believed that the Sun revolved around the Earth. Still today, you and I use words like *sunrise* and *sunset* to describe what appears to us as the Sun's daily movement *around* planet Earth. But, of course, it is Earth, not the Sun, that is moving. At the end of the day Earth literally rolls away from the Sun, giving the mistaken impression that the Sun is setting. Here is a trick that you can use to correct your perception. The next time you are watching a sunset, imagine pinning the Sun to the wall of the sky in such a way that the Sun remains stationary. Then, watch as the horizon moves up to meet the Sun. Seen in this way, you will be able, with some practice, to experience that the place on Earth where you are standing is rolling away from the Sun, moving you into darkness.[3]

large, dim ones tiny. The possibility that the dimness of stars might owe to their remoteness in space was, for the most part, not seriously entertained. Indeed, early astronomers may well have imagined that the stars were only a few thousand miles away from Earth. Living in this mental construct, the only object of significant size in the universe would have been Earth, which, of course, would explain why Earth stood at the center, with everything else circling obediently around it.[4]

But then along came Copernicus, Galileo, Kepler, and others. Using empirical observations of the heavens based on telescopes and more sophisticated mathematics, these investigators blew the lid off the celestial spheres hypothesis, positing that it was the Sun, not Earth, that was at the center of the universe; and that the Sun was not close by, but tens of millions of miles away. For us this all seems so obvious, but put yourself back five hundred years. One evening:

> You go out under the stars to consider these ideas. You know the stars well for they are your calendar and nighttime clock. You stand at the center of your familiar universe, looking up at the well-known, nearby stars, when suddenly—it can happen no other way—the celestial sphere shatters and the stars are hurled at varying, unknown distances.[5]

In that moment you grasp, for the first time, that a star's brightness or dimness is, to a significant degree, the result of how far away it is. You stand awestruck as you consider that there might be stars so far away that they are too faint to see. Your world is no longer small: the universe may be too big to see, and you are no longer at the center. Feel the dissonance, as your old worldview—the old story—collapses, but also feel possible excitement as you are beckoned to construct a new worldview.

A New Story

Just as people in the 1600s resisted the new knowledge presented to them by science, so it is that today we sometimes find it difficult to embrace new scientific discoveries. For example, nowadays we accept that planet Earth revolves around the Sun (even though the opposite appears to be the case); but most of us struggle to comprehend that our Sun is just one of tens of billions of stars in the Milky Way galaxy and that our galaxy is just one of the hundred-billion-plus galaxies that make up the known universe.

It has been humbling for humankind to acknowledge that Earth is not at the center of the universe. Indeed, it turns out that planet Earth is a small-ish planet circling a smallish star (our Sun) out toward the edge of the Milky

ONE HUNDRED BILLION GALAXIES!

Suppose you were assigned the task of naming all of the galaxies in the known universe! Suppose further that you were really fast at naming—able to name a new galaxy every second—and that you were able to perform your naming task day and night without sleeping. Even with these advantages, it would still take you more than three thousand years to complete your naming project!

Way galaxy in a universe seething with more than a hundred billion other galaxies. And, yet, in a fascinating way, our galaxy—all galaxies, really—is at the center of things because, as we now know, the universe is omnicentric, meaning that there is no one single center. Indeed, not long ago, astronomers discovered, to their astonishment, that all the known galaxies (or galaxy clusters) are moving away from each other. One way to visualize this is to think of a loaf of raisin bread baking in the oven. Let each raisin be a galaxy. Place yourself on one of the raisins. As the bread bakes and expands and as you look around, you will see all the other raisins (galaxies) moving away from you. You would perceive the same thing if you were positioned on any other raisin, as well. Thus, no matter which raisin (galaxy) you are on, you will see your galaxy as the center.[6]

Not only are the galaxies expanding apart from each other, but the farther away they are, the faster they are speeding apart. So it is that galaxies that are twice as far apart are separating two times faster, and those that are ten times farther apart are racing away ten times faster. Space, literally, rushes into existence as the galaxies separate from one another.

The Big Bang

Scientists have been able to trace this process of galaxy expansion *backward* to a hypothesized beginning point—the birth of the universe!—estimated to have occurred some fourteen billion years ago. Think of it as running a movie backward and employing the laws of physics to determine when things began. Imagine seeing this movie—that is, watching the myriad galaxies of today's universe reverse their outward expansion and begin to draw together, eventually merging into a hot, high-density gas. And with further contraction and increasingly stupendous temperatures, imagine all atomic mass converted to pure energy as the entire universe collapses into an infinitely small, infinitely hot, mathematical point—the so-called singularity that gave rise to the big bang fourteen billion years ago.[7]

A FOURTEEN-BILLION-YEAR-OLD UNIVERSE!

The notion of fourteen billion years, just like the idea of one hundred billion galaxies, is very tough to grasp. So see if this helps: Imagine that a distance of one millimeter is equivalent to one year. Thus, one meter (composed of 1,000 millimeters) would equal one thousand years, and one kilometer (composed of 1,000 meters), a million years. Using this scale, the entire fourteen-billion-year history of the universe would roughly span a distance from the center of Tokyo (i.e., the beginning of the universe) to Washington, DC (today). Now, imagine yourself preparing to walk back to the beginning of time. You are standing at the base of the Washington Monument. Your first step (about a half-meter) would take you back five hundred years to the time when the Spanish first sailed to America. Four more steps and you would be back to the time of Socrates. Trekking through northern Virginia you would be back in the age of the dinosaurs (65 million years ago). To arrive at the time when the Earth and Sun were formed (roughly 4.6 billion years ago), you would have to walk all the way to the California coast. And if you were truly earnest in traveling to the beginning of time (back to the big bang), you would still have to make your way across the immense expanse of the Pacific Ocean, all the way to Tokyo.[8]

Scientists don't simply "run the movie backward" and then accept that the universe began from a big bang fourteen billion years ago. They check for other evidence. For example, here on Earth they do experiments using high-energy particle accelerators to understand what happens to matter at excruciatingly high temperatures. Then they use this information to calculate the kind of universe that would have emerged from the hypothesized big bang. These calculations predict that the matter formed from the big bang should have consisted of hydrogen and helium in a ratio of three to one, which is, in fact, the makeup of the universe today as determined by astrophysicists.[9] Furthermore, if there really was a big bang, it stands to reason that the faint light from that ancient cosmic explosion should still be detectable. Although this radiation wouldn't be visible to the naked eye, scientists reasoned that it could be measured with infrared and radio telescopes. And, indeed, in 1989, the satellite dubbed COBE (Cosmic Background Explorer) measured this radiation with great precision and found that its temperature and behavior matched that predicted by big-bang theory. Here's the punch line: The big bang isn't lost far back in time; today Planet Earth receives light that was emitted when the universe was first forming. The big bang is all around us and in us.[10]

The Big Bang in Us?

I remember learning in high school that our bodies are 70 percent water. In other words, without our skeletons to hold us upright, we'd be watery bags sloshing around on the ground. Another fact I absorbed—this one not very interesting at the time—was that a water molecule is composed of two atoms of hydrogen plus one atom of oxygen. It was only much later, when I began to wonder about the actual origins of water's two constituent atoms, that I learned something that I find truly astonishing. Ready? Drum roll! Remarkably, all the hydrogen now on Earth—all the hydrogen in your body—was formed some fourteen billion years ago when the universe exploded into being. Now get this: Hydrogen was *only* created in that primordial flaring forth—the big bang—and never again since that time. And it was hydrogen that gave rise to the galaxies and first stars.

Consider the implications of this the next time you prepare to drink a glass of water. Given that all the hydrogen in water is some fourteen billion years old and that hydrogen is the most abundant atom in your body, how old are you? Your human age might be fifteen or fifty or eighty but the stuff of you is fourteen billion years old; you've been here since the very beginning.

What Lies Ahead

I have skipped over a huge question—namely: What caused the big bang in the first place? When this question is posed to physicists, they explain that it is theoretically possible for something to materialize out of a vacuum—out of nothingness! Hmmm. According to MIT physicist Alan Guth, a speck of matter one-billionth the size of a proton may have bubbled into being and created a repulsive gravitational field—a so-called false vacuum—so strong that it exploded into our universe. Apparently, this is theoretically possible. Science writer Brad Lemley explains:

> According to Einstein's theory of relativity, the energy of a gravitational field is negative. The energy of matter, however, is positive. . . . Calculations totaling up all the matter and all the gravity in the observable universe indicate that the two values seem to precisely counterbalance. All matter plus all gravity equals zero. So the universe could come from [essentially] nothing because it is fundamentally, nothing.[11]

If you think this scenario sounds weird, you are not alone. Even the scientists involved in this research acknowledge that their findings are disorientating and difficult to fathom.

The question of how the universe will end has also occupied the scientific community in recent decades. There appear to be two possibilities—either the universe will continue expanding forever or it will eventually reverse direction and collapse back on itself. It all depends on the interaction of two forces—the speed at which the galaxies are rushing apart and the mutual attraction of gravity that counteracts their outward rush. The situation can be compared to that of a rocket ship leaving Earth. The rocket must achieve a velocity of at least twenty-five thousand miles per hour to escape Earth's gravitational pull. "Like a rocket engine, the big bang hurled the galaxies outward; gravity is pulling them back together. Do the galaxies have sufficient speed to overcome their mutual attraction? Do they possess the necessary escape velocity?"[12]

If the universe continues to expand (as most scientific evidence now indicates), it will end with a whimper tens of billions of years in the future, when there is no more energy for the birthing of new stars. If, on the other hand, the galaxies lack sufficient speed to overcome their mutual attraction, the universe will eventually begin to contract, ending just as it began, but instead of a big bang, there will be a big crunch. Either way, the universe has a finite lifetime—or does it? Cosmologists Andrei Linde and Alexander Vilenkin, operating on the assumptions of Guth's inflation theory (above), remind us that it is theoretically possible to have an infinite number of inflating universes.[13] How science delights us with its mysteries!

FOUNDATION 1.2: ORIGINS AND WORKINGS OF THE SOLAR SYSTEM

The famous astronomer and cosmologist Carl Sagan observed that we are all stardust. It is only by studying how stars, such as our Sun, and planets, such as Earth, are formed that we can appreciate the literal truth in Sagan's words.

Formation of Stars and Planets

Our Sun came into being roughly 4.6 billion years ago, and it may surprise you to learn that it owes its existence to the *death* of pre-existing stars that were formed when the universe was still relatively young. As those so-called first-generation stars burned up all the hydrogen fuel in their cores, they generated heavier elements like carbon, nitrogen, oxygen, calcium, and silicon that were released in a matrix of dust and gas. And in cases where dying stars were really large—that is, many times larger than our

Sun—their death was marked by monstrous explosions called "supernovae" that released very heavy elements like iron, nickel, and gold into space. Though shocking, without star death, new stars, like our Sun, and planets, like our Earth, would not exist.

Use your imagination to go back five billion years in time and picture the gas and dust generated by supernovae gathering along weak lines of magnetic force and then gradually collecting in whirlpools and eddies; watch as the atoms begin to clump and form a common center. Sense the enormous attractions and pressures that lead the hydrogen atoms to fuse into helium atoms, provoking massive outbursts of energy that cause the core to ignite, forming a star. First there was a cloud of hydrogen atoms; then, because of the immensity of the undertaking and the forces that come into play, our Sun was born.

And what of the formation of Earth? How did that come about? Here is how Brian Swimme and Mary Evelyn Tucker describe it in their lovely book *Journey of the Universe*:

> In the beginning our infant sun was completely surrounded by hydrogen, carbon, silicon, and other elements disbursed by the supernova explosions. As they drifted through space these elements would brush against each other and begin to cohere into tiny balls of dust. Over millions of years these "planetesimals" continued accreting and growing until they were the size of boulders and then as large as mountains. Not all collisions resulted in larger bodies. Many were so violent that they tore both bodies apart. But over millions of years these planetesimals continued to absorb all the loose material circling about. Our solar system, with its eight planets, its band of asteroids, and its one infant sun, slowly came into being.
>
> It is remarkable to realize that over immense spans of time stellar dust became planet. In the earliest time of the universe this stellar dust did not even exist because the elements had not yet been formed by the stars. Yet hidden in this cosmic dust was the immense potentiality for bringing forth mountains and rivers, oyster shells and blue butterflies.[14]

As for you and me—every part of our bodies contains this stellar dust. Yes, we are stardust through and through. It was through the bestowal of the gifts of myriad dying stars that blue whales, prairies, clouds, birch trees, butterflies, computers, strawberry sundaes—*everything!*—on Earth has become possible.

The Workings of the Sun

It is easy to take the Sun for granted. It has been burning brightly almost five billion years, and astrophysicists calculate that it still has five billion years of fuel left. Only in the twentieth century did scientists come to understand how the Sun works. It is an extraordinary story that offers a window into $E = mc^2$—Einstein's famous equation that describes how mass is converted to energy. According to this equation, the amount of energy (E) produced by the Sun is related to solar mass (m) multiplied by the speed of light squared (c^2). What a remarkably simple equation—two things multiplied together to give an answer! And one of those two things, the speed of light (c), is already known—186,000 miles per second. The only thing missing to solve the equation is solar mass (m). Yes, the Sun gives planet Earth energy by sacrificing part of its mass. This occurs at the Sun's center where stupendous temperatures cause protons to fuse together. Normally, it is hard to get two protons to fuse because their positive charges repel each other ferociously. Here on Earth, for example, this union only happens either in enormously expensive particle accelerators or in the fury of atomic bomb detonations.[15] But the immense size of stars, like our sun, creates the massive gravitational forces and associated temperatures that allow for nuclear fusion. Now here is the key part: As protons fuse at the Sun's core, they lose a small amount—about 1 percent—of their mass. This lost mass is the *m* in $E = mc^2$. Though the amount of lost mass is tiny, when it is multiplied by the speed of light squared (i.e., by approximately 35 billion) the result is large.

The energy released from proton fusion at the Sun's core percolates upward until it reaches the Sun's roiling surface where it is hurled into space as heat-light. Indeed, eight billion pounds of the Sun's proton mass is lost each second. That's right, eight billion pounds! Of this vast quantity, planet Earth only intercepts five pounds' worth. Think of it as five pounds of vanished solar mass arriving to Earth as radiant energy each second—lighting up and powering Earth. Five pounds of solar mass doesn't sound like much until we understand $E = mc^2$. In fact, the amount of solar energy intercepted by Earth in just one hour is greater than the total amount of fossil fuel energy humankind uses in one year.[16]

If it weren't for the Sun's warming radiation, temperatures on Earth would be hundreds of degrees below zero—far too cold for life to exist. And without the Sun's rays, there would be no evaporation of water from oceans and leaf surfaces and that would mean no rainfall, no hydrological cycle! Nor would there be photosynthesis. Yes, the Sun is, in effect, a giant *generator* that powers the winds, the water cycle, photosynthesis, and, ultimately, our very bodies. No wonder so many ancient peoples regarded the Sun with a sense of awe and reverence!

Recalling the entirety of this epic story, we see that our solar system, coalescing into existence from cosmic gas and dust generated by solar winds and star explosions, has now given birth to us. Incredibly, our very bodies, our cells, the air we breathe, are the work of the heavens—gifts from the stars!

The Origins of Life on Earth

By remarkable good fortune, Earth offers conditions propitious for life as a result of its distance from the Sun and its size. For example, Earth is not so close to the Sun that its surface is scalded, as is the case with Mercury; nor is Earth so far from the Sun that its life-giving water is perpetually frozen, as is the case with Jupiter. Likewise, Earth is not so large that its gravity holds a dense blanket of atmospheric gases that filter out sunlight, nor is it so small that it fails to hold onto a life-sustaining atmosphere.[17]

Although Earth is well situated, its environment at birth—a fiery magma ball—was hostile to life. With time, however, a crust formed over Earth's molten core, much like the thin surface layer that forms on pudding as it cools. During the cooling process, gases were released, primarily carbon dioxide, nitrogen, and water vapor. Earth's early atmosphere then acted as a thick heat-trapping blanket. This was a steamy, gaseous time for Earth; and as Earth slowly cooled over millions of years, vapor condensed into clouds and rain fell . . . and fell . . . and fell—indeed, fell for so long that oceans were formed.[18]

Water served as a kind of planetary womb, providing the necessary medium to nurture life; and it was probably in Earth's primeval seas that life first appeared, but nobody knows how. Was it simply by chance? Was there some as yet undiscovered self-organizing principle involved? We simply don't know.

One science-based theory of life's origins, known as "panspermia," holds that life didn't arise on Earth at all, but somewhere else—out in space. According to this view, the primitive seeds of life are adrift throughout our galaxy; and, by chance, they occasionally colonize a planet like ours, arriving via meteorites or comets.

Another theory proposes that the first complex organic molecules on Earth were produced with energy contributed by lightning and/or by the release of heat from deep-sea vents. To test the efficacy of this theory, scientists placed water—to represent the early oceans—in closed glass containers and added the gases that were presumably present in Earth's primeval atmosphere. Then they applied heat and electric sparks to simulate lightning, and in so doing they were able to generate protein building blocks, fatty acids, and urea (a molecule common to many life processes); but, of course, a collection of organic molecules, by itself, still doesn't constitute life.[19]

Somehow, the molecules for life must have become encased in a cell. But how? It is likely that among the many compounds in Earth's early chemical soup, there were greasy compounds (i.e., lipids) that formed bubbles (cells) when mixed with water. Microbiologist Lynn Margulis sketches out how life's first cells might have come into existence:

> Cell-like, membrane enclosures form as naturally as bubbles when oil is shaken with water. In the earliest days of the still lifeless Earth, such bubble enclosures separated inside from outside. Pre-life, with a suitable source of energy inside a greasy membrane, grew chemically complex. . . . These lipid bags grew and developed self-maintenance. . . . After a great deal of metabolic evolution . . . some [cell prototypes] . . . acquired the ability to replicate more or less accurately.[20]

Of course, this is pure speculation, but it forms the basis of a hypothesis that scientists can investigate.

This science-based perspective need not be seen as, necessarily, in conflict with creation stories coming from the world's different cultural and religious traditions. All these stories—both religious and scientific—represent a yearning for meaning and understanding; they are simply different ways of fathoming the same mysteries, and at their core they share some remarkable commonalties. For example, in Genesis in the Christian Bible we read, "And God said, 'Let there be light and there was light,'"[21] which coincides neatly with the *big bang* story that the universe began in an explosion of light.

The History of Life on Earth

Earth's first organisms, originating approximately 3.9 billion years ago, were primitive forms of bacteria. It has only been in the last 10 percent of Earth's history—starting just six hundred million years ago—that more complex life forms have appeared—such as mosses, ferns, fungi, crustaceans, amphibians, reptiles, insects, and flowering plants. And it wasn't until just two hundred million years ago that, according to the fossil record, the first mammals came into existence. Our human body form—composed of head, trunk, and arms and legs—is present in all mammals as well as in some earlier life forms such as reptiles and amphibians. Yes, amphibians and reptiles have a head, trunk, and four limbs just like us. We are connected to them by this basic plan that was an invention of Earth some four hundred million years ago.

At present, the diversity of life on Earth is extraordinary. The total number of species is estimated at more than ten million, although only one and a half million species have been formally classified. Many of Earth's species are too small to be seen. For example, using just the naked eye and searching above ground, one could locate several hundred different species in a patch of forest in North America—worms, snakes, frogs, caterpillars, rodents, herbs, trees, mosses, and so forth. Several thousand additional species—such as mites, tiny crustaceans, fleas, worms, algae, bacteria, and fungi—could be located with the assistance of a microscope. Indeed, an invisible world of living organisms is located in the soil, on the bark of trees, and on the surfaces of leaves. This process of searching for species in ecosystems is akin to the search for stars in the night sky using high-power telescopes. Consider:

> Most of the stars in the sky cannot be seen with the naked eye, but with the help of sophisticated technology, millions of stars appear in the cosmos. In biology, it is microscopes that reveal the otherwise invisible objects. In this case they are species and the "cosmos" is the Earth's myriad environments and habitats.[22]

Although humans are latecomers in this evolutionary saga, our origins actually extend all the way back to the appearance of single-celled, bacterial life on Earth. Indeed, our earliest ancestors were not primitive primates; they were bacteria! This might sound crazy, but how could it be otherwise? Life arose, evolved, and complexified until eventually our hominid form arose; but it all began with bacteria. We evolved from bacteria; we stand on their shoulders. And it's even possible, with a little imagination, to trace our ancestry further back still. After all, we wouldn't be here were it not for stars that expired long ago, creating, in their dying, the primordial materials that gave rise to Earth and with it, us. Yes, the stars *are* our ancestors!

Life on Other Planets?

Given the abundance of life on Earth and the immensity of the known universe, what is the possibility of life on other planets? This question is fun to think about. Consider this possibility just for our Milky Way galaxy, one of an estimated hundred billion galaxies that make up the known universe.

Table 1.1.   The History of the Universe from Fourteen Billion Years Ago to the Present

Years before Present	Cosmic/Evolutionary Event
14 billion years	Big bang—the primordial flaring forth of space, time, and matter.
13.99 billion years	Hydrogen and helium, the first atomic "life," coming into existence.
13 billion years	Galaxies and stars forming from the hydrogen and helium issuing from the big bang.
4.6 billion years	Stardust and gases from the dissolution of first-generation stars offering raw material for the formation of our solar system.
3.9 billion years	Earth coming to life; bacteria appearing in Earth's primitive seas.
3.6 billion years	Bacteria inventing photosynthesis whereby they capture light energy from the Sun.
2 billion years	Two distinct forms of life merging to form the first nucleated cells.
1 billion years	Single-celled organisms reproducing sexually by combining genetic material.
600 million years	The first multicellular organisms appearing.
500 million years	The first organisms with backbones moving through the oceans.
400 million years	Life forms leaving water; worms, mollusks, and crustaceans breathing air for the first time; algae and fungi venturing ashore.
300 million years	The first forests appearing; reptiles and amphibians flourishing on land.
200 million years	Dinosaurs diversifying in size, form, and habits; mammals appearing and birds budding off from dinosaurs.
100 million years	Flowers evolving and insects diversifying to feed on nectar and pollen.
65 million years	Dinosaurs disappearing; mammals flourishing: whales, rodents, antelopes, cats, horses, elephants, camels, bears, baboons appearing.
4 million years	Human ancestors leaving the forest and standing up to walk on two legs.
2 million years	Early humans using their hands and heads to shape tools.
100,000 years	Modern humans emerging; language, clothing, shelter, fire, musical instruments, art, and religion appearing.
10,000 years	Humans learning to cultivate plants and domesticate animals.
100 years	Humans discovering that they live in an expanding universe.
50 years	Humans traveling to outer space and looking back to see their home, planet Earth, for the first time.
Today	Humans realizing that they are the result of the universe's ongoing creative expression.

Source: Sidney Liebes, Elisabet Sahtouris, and Brian Swimme, *A Walk through Time* (New York: John Wiley and Sons, 1998); Brian Swimme and Thomas Berry, *The Universe Story* (San Francisco: Harper San Francisco, 1992).

Within this galaxy, our Sun is one of one hundred billion–plus stars. That's one hundred billion other suns, each with the potential to have solar systems of its own; but, to be conservative, let's imagine that only one half of those suns in our galaxy actually have planets. This leaves us with roughly fifty billion suns, each with planets, in the Milky Way galaxy.

However, planets with the characteristics that make them candidates for life—just the right distance from their sun and just the right size for the formation of a life-sustaining atmosphere—might be relatively rare. Thus, let's be conservative again and grant that only one in every five hundred of those fifty billion suns in our galaxy has a planet with the basic characteristics required for life as we know it. This still leaves us with roughly a hundred million planets with the potential for life support in just our galaxy.

With the possibility of extraterrestrial life seemingly all around us, perhaps a more appropriate question is: Why haven't we had any visitors? Or if we are having visitations, why aren't they more common? The reason, in large part, may be the extreme size of our galaxy. The average distance between stars, considering all stars, is five hundred light years. And how big is that? Well, one light year is the distance that light, moving at the speed of 186,000 miles per second, travels in one year. And how far is that? Let's do the math: 186,000 miles/second × 60 seconds/minute × 60 minutes/hour × 24 hours/day × 365 days/year = 5,865,696,000,000 miles/year. Yes, one light year is approximately six trillion miles—a colossal distance—and the average distance between stars is not one light year, but five hundred light years! And, of course, the average distance between that particular subset of stars with potentially life-supporting planets would be even greater. In addition to these vast distances in space, we also have "distance" in time— that is, possible civilizations would be scattered in time. To grasp this concept, consider that our solar system has been around for almost five billion years, but intelligent life (e.g., us), capable of reaching out into space, has been in existence for only a very, very tiny fraction of that time. Thus, at any one time, we may only have a few thousand stars in our galaxy simultaneously supporting a planet with life; and the average distance between these stars, given the size of the Milky Way, would likely be on the order of thousands of light years. None of this rules out the possibility that there could be extraterrestrial beings in our galaxy that are unimaginably more complex, intelligent, and marvelous than we are—beings that perhaps have already checked us out and concluded that we are not very interesting given our primitive state. It is a humbling thought.

STEPPING BACK TO SEE THE BIG PICTURE

My body and the universe come from the same source, obey the same rhythms, flash with the same storms of electromagnetic activity. My body can't afford to argue over who created the universe. Every cell would disappear the second it stopped creating itself. So it must be that the universe is living and breathing through me. I am an expression of everything in existence.

—Deepak Chopra[23]

Humans, since our earliest days on Earth, have puzzled over and tried to make sense of the cosmos through story. In our time, this process continues as cosmologists take recent scientific knowledge about the origins of the universe, stars, planets, and life and weave this into a new story. It is noteworthy that our universe is structured by a force that we call "gravity"—a force of attraction that keeps objects connected or in relationship to one another. This seems to have been true from the very beginning of time. For example, current mathematical models of the early universe assert that elementary particles (i.e., quarks, leptons, Higgs boson) came into being at the moment of the big bang. Then these particles quickly morphed into protons and neutrons and began to bond together in stable relationships. Yes, from the very first moments, it appears that the universe moved toward creating relationships:

> Things could have been different. We can theorize about a different kind of universe, a universe that would have taken the form of disconnected particles, a universe that would never have formed bonded relationships. . . . But in our observable universe, various forms of bonding are inescapable. . . . Bonding is at the heart of matter.[24]

I find it reassuring to live in a universe grounded in relationship. Likewise, when I learn that our Sun gives five pounds of itself each second to Earth— so that we can all live—I feel a mix of wonderment and gratitude, recognizing that a forerunner of human generosity is stellar generosity.[25] And when I grasp that the most common element of my body, hydrogen, originated fourteen billion years ago at the birth of our universe, I stand a little taller knowing that my roots are ancient indeed.

In addition, when I learn that violent cosmic explosions (i.e., supernovae) catalyzed the creation of our solar system, I realize that I live in a profoundly dynamic universe. Stars and life are snuffed out in fits of destruction and ushered forth in creative outbursts. This interpretation helps me to understand my own life. I, too, have times of stasis and times of upheaval, and often my creativity is triggered by upheaval.

The most remarkable thing about the universe story is that it is not over. Indeed, we live in the midst of an unfolding universe, and, as such, we are participants in an epic story. It all began with the creation of hydrogen in the big bang; now, fourteen billion years later, hydrogen has differentiated and complexified into corn plants, butterflies, and gazelles, not to mention us! In effect, energetic and gravitational forces have acted on hydrogen atoms over billions of years, transforming it into all that is. This is the astonishing story of creation that has come forth in the twenty-first century.

IS THE UNIVERSE ALIVE?

Elisabet Sahtouris, in her book *EarthDance*, invites readers to consider that the universe as a whole is alive when she writes: "If we agree that nature is not mechanical but organic, why should we not understand the energetic motion of the very first whirling shapes in the early universe as the first stirring of [a] self-organizing process leading to living organisms? The spiraling pre-galactic clouds, composed of spiraling atoms, held themselves together, drew in more matter-energy from their surroundings, built it into themselves, and lost energy again to their surroundings. In this process of energy exchange they evolved into new, more complicated forms. By the time we get to galaxies and to fully formed stars within them, the dance toward life has become quite complicated already. Our living Earth is likely no more a freak accident than is the seedling that grows or the frog egg that matures. All are the inevitable result of right compositions and conditions."[26]

In this new story, the universe is not a thing at all, nor is it a place; rather, it is a process of becoming—a dazzling unfolding. Indeed, the universe has an innate tendency to self-organize and thereby to become more complex, more highly organized, more conscious, and more fully alive, over time.

If you believe in God (or gods), revelations from this new story need not undermine or contradict your faith. Indeed, these new revelations from science could just as easily serve to deepen and/or broaden your understanding of the divine. For example, in this new story, rather than understanding God as a *being*—a noun—God might be understood as more akin to a verb—an action or process—that is, the beingness that infuses all things—human being, animal being, plant being, atmospheric being, geologic being. In this way, the new story abandons the old belief that the world was made for humans, putting in its place the belief that we, humans, were made for the world. In other words, it is *not* the world that belongs to us, but we who belong to the world. This shift presents a profound challenge to human identity.[27]

In sum, science and religion, rather than being mutually antagonistic have the potential to become mutually enriching. For example, you could use the new scientific information about the nature of the universe to make inferences about the nature of God (or your placeholder for God). How? By simply asking what kind of Creator would invent a universe like ours.

- What kind of creator would place coupling and relationship at the center the universe?

- What kind of creator would design a universe that is wondrous and mysterious beyond our wildest imaginings—for example, a universe that, to our eyes, is both imperceptibly large *and* imperceptibly small?
- What kind of creator would fashion a universe where destruction leads to rebirth and transformation?
- What kind of creator would manifest a universe that is not fixed or static but dynamic—in a constant and active process of moving toward more complexity, greater diversity, and more intimacy?

As I playfully explore these questions, words like *loving, creative, generative, spontaneous, whimsical, wise, benevolent* arise as descriptors of the unnamable force underlying creation. The point, of course, is that the universe, itself, can become a profound source of revelation.[28]

In the end, while science can tell us some of what happened and can sometimes offer powerful explanations for how it happened, first-order mysteries may always lie beyond its reach.

WRAP-UP: EXPANDING ECOLOGICAL CONSCIOUSNESS

> It's all a question of story. We are in trouble just now because we do not have a good story. We are in between stories. The old story, the account of how the world came to be and how we fit into it, is no longer effective. Yet we have not learned [a] new story. . . . We need a story that will educate us, a story that will heal, guide, and discipline us.
>
> —Thomas Berry[29]

Now that you have completed this chapter here's a culminating question for you to ponder: *Where is the universe? Point to it.* Given this prompt, most people point away from themselves, out into space. It seldom occurs to us to point inward because we are not accustomed to seeing ourselves as an integral part of the universe—that is, we have little awareness that the universe is inside of us. But, of course, when we look up at the stars, we are, in effect, looking at ourselves. This is not figuratively true but literally true. The calcium of our bones, the iron of our blood, the carbon of our tissues all had their origins in the belly of stars—stars that in their dying, seeded the heavens with the elements of life. Both your body and the universe are composed of the same molecules, come from the same source, obey the same rhythms.

For my part, I admit that I often still live in the default consciousness of the 1500s—that is, in that *old story* of a tiny, static universe with Earth at the center. Yes, I struggle to experience myself as a citizen of the Milky Way, dwelling in an expanding universe. Instead, on a daily basis, my consciousness is usually tethered to the physical spaces I occupy. Living in this way, with my focus limited to my daily life concerns, is a bit like perceiving the world through a straw, only experiencing a sliver of reality. I know this because in those special moments when my awareness expands—such as gazing into the inky depths of the night sky from a mountain top—my experience of being alive is no longer small and pinched but expansive and ecstatic, as I remember that life is a glorious and miraculous gift to be partaken of with gratitude and, yes, reverence.

In sum, human consciousness will expand as we develop the capacity to see our lives in the context of a wondrous, unfolding universe. But this expansion doesn't come by simply reading a book chapter; it requires that we actively cultivate a relationship with our solar system, with our home galaxy, the Milky Way, and with the universe that has brought us into being. In short, the call is to wake up to our embeddedness in something far grander than our minuscule selves and in this way to experience a genuine *homecoming*.

APPLICATIONS AND PRACTICES: CULTIVATING DISCOVERY

> Just as the Milky Way is the universe in the form of a galaxy, and an orchid is the universe in the form of a flower, we are the universe in the form of a human. And every time we are drawn to look up into the night sky and reflect on the awesome beauty of the universe, we are actually the universe reflecting on itself. And this changes everything.[30]

Discovering Our Relationship to the Universe

Spending most of our time indoors, as so many of us now do, we easily forget that we dwell in a universe. Fortunately, we can all engage in an age-old practice—stargazing—to acquaint ourselves with our larger home. It is not hard to do. On a cloud-free night, pack a blanket and a thermos; and with some friends, head out to a quiet spot in the country or the mountains, far from the glare of city lights. Then, spread out your blanket, and lie down and gaze up at the night sky. In so doing, you will be looking at the Milky Way galaxy from within its boundaries.

If you could somehow remove yourself from the Milky Way and look down on it from outside, it would appear like a gigantic Frisbee—one hundred

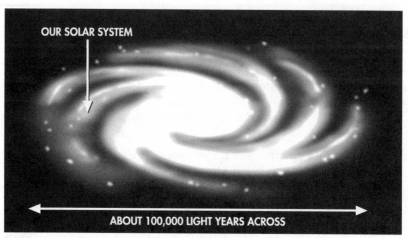

OUR SOLAR SYSTEM

ABOUT 100,000 LIGHT YEARS ACROSS

Figure 1.1.   A representation of the Milky Way galaxy.

thousand light years across—with a bulge at the center (figure 1.1).[31] The Milky Way galaxy is not out there separate from you; you live within this great wheel of stars. As you gaze skyward, you will see stars in all directions; but when your gaze is straight across the plane of the Milky Way, the commingling of light from myriad distant stars will cause you to see a shimmering milky path.

Now, as you lie there gazing into your home galaxy, imagine yourself whirling on your multiple journeys. First, recall that the Milky Way (with you included) is still racing outward from the big bang. Next, consider that Earth, in addition to spinning on its axis each day and orbiting around the Sun each year, is involved in a grand galactic rotation—whereby it spirals around the vast Milky Way galaxy once every 225 million years.

And here is another thing to contemplate: As you lie on your back, it is natural to assume that you are looking up at the stars, but cosmologist Brian Swimme reminds us that "up" is just a cultural construct. Neither Earth nor the Milky Way have an up or a down. Indeed, when you stand on Earth's surface, you are not standing up; rather, you are sticking out into space. So, as you lie on your back, instead of thinking of yourself as looking up, picture it so that you are on the underside of Earth looking down into the blackness of the night sky. It may take a while, but eventually you will experience all the stars as way down there below you; and you will be surprised that you are not falling down there to join them.

You don't fall, of course, because Earth's gravitational pull holds you. It is not your weight, but the Earth's hold that suspends you above the stars. If

Earth's gravitational embrace were to suddenly vanish, you would descend into the dark chasm of stars below:

> As you lie there feeling yourself hovering within this gravitational bond while peering down at the billions of stars drifting in the infinite chasm of space, you will have entered an experience of the universe that is not just human and not just biological. You will have entered a relationship from a galactic perspective, becoming for a moment a part of the Milky Way Galaxy experiencing what it is like to be the Milky Way Galaxy.[32]

Sky watching can be just a movie that you go to once and say "Been there!" but it can also be a practice. Sky watchers know that no two nights are the same. Atmospheric conditions, hour of night, time of year, and physical location all affect what can be seen. Binoculars can enhance the experience. But why stop there? Chet Raymo suggests that the next time you go stargazing you take along a CD player and a recording of Joseph Haydn's *Creation Oratorio*. Once you are comfortably settled, close your eyes and breathe yourself into a relaxed state. Then, with your eyes still closed, press play:

> Silence. A C-minor chord, somber, out of nowhere. Followed by fragments of music. Clarinet. Oboe. A trumpet note. A stroke of timpani. A prelude of shadowy notes and thrusting chords, by which Haydn meant to represent the darkness and chaos that preceded the creation of the world. . . . Listen now, eyes closed, as the music descends into hushed silence. . . . Open your eyes! A brilliant fortissimo C-major chord! A sunburst of sound. Radiant. Dispelling darkness. A universe blazes into existence, arching from horizon to horizon. Stars. Planets. The luminous river of the Milky Way. As you open your eyes to Haydn's fortissimo chord and to the (almost) forgotten glory of a truly dark starry night, you will feel that you have been a witness to the big bang.[33]

Practices that befriend the night sky can help us to experience our lives in a larger context. At present, many of us are more likely to be sitting in front of the television at night than under a star-strewn sky. The state of consciousness engendered by television versus that of the night sky is strikingly different: One invites us to dwell in a world of human artifice; the other calls forth humility, inviting us to experience what it is like to belong to something much grander than ourselves.

Discovering Your Relationship to the Sun and Seasons

We have all been conditioned to understand and experience time as a vector, defined by the past, present, and future—what the Greeks called "chronos." But, just for fun, try to imagine the experience of time, absent any concept of past or future, absent calendars and clocks. The Greeks referred to this present-moment experience of time as "kairos," as distinct from chronos.[34]

It is possible to cultivate *kairos* through simple daily observances. For example, go outside first thing in the morning and simply experience the day: the temperature and movement of the air; the condition of the sky; the movements and sounds of the creatures that are co-inhabitants of your home place. And as your day unfolds, pay attention to the changing position of the Sun and the changing qualities of light as Earth turns on its axis. Note the first signs of a change in the weather. When night comes, step outside again and pause to pay attention to the air, its humidity and movement. In this way, you can resynchronize your biological clock to the rhythms of Earth.

For time immemorial humans have signaled their relationship to the Sun and seasons through celebration and ritual. In our time, we would be wise to do the same. A good place to start is with children—using simple ceremonies that help them appreciate the Sun's gifts. In this vein, Swimme asks us to imagine awakening our children a few minutes before dawn for a Sun ceremony:

> As they're yawning away the last of their sleep and as the Earth slowly rotates back into the great cone of light from the Sun, they will hear the story of the Sun's gift. How five billion years ago the hydrogen atoms, created at the birth of the universe, came together to form our great Sun that now pours out this same primordial energy and has done so from the beginning of time. How some of this sunlight is gathered up by Earth to swim in the oceans and to sing in the forests. And how some . . . has been drawn into the human venture, so that human beings themselves are able to stand here, are able to yawn, are able to think, only because coursing through their blood lines are molecules energized by the Sun.[35]

Rituals like this one don't require any significant planning. All we have to do is create a space for them to happen and trust that we will know what to do.

QUESTIONS FOR REFLECTION

- What does it mean to you to live in a universe?
- How does this new understanding of the universe—not as a static entity, but as a dynamic happening—change your understanding of yourself? Your past? Your future?

- What might you do on a daily basis to acknowledge your utter dependence on the Sun's benevolence?
- The universe is bathed in mystery. . . . What are the mysteries in your life?
- How does the so-called new story affect your understanding of the divine or, if you prefer, the great mystery?
- What could you do to cultivate more kairos (and less chronos) in your life? And why might it be worthwhile to do so?

Responding to Questions: The Power of Reflection

Throughout this book, I offer *questions for reflection* at the end of each chapter. I pointedly don't call these "Reaction Questions," because our reactions are simply conditioned responses and, as such, they require no thought and lead to no new understanding. Reflection, on the other hand, requires genuine thought. Indeed, to reflect on a question demands that we sit with it, ponder it, and, ultimately, that we open to the possibility that the question, if taken seriously, can lead us to new understandings. Here is a four-step exercise for exploring this distinction between *reflecting* and *reacting.*

Step 1. Pick a question from the above list that is intriguing to you. For illustrative purposes, let's suppose that you choose the question "What does it mean to you to live in a universe?" Now, offer your first *reactions* to this question. No reflection, just reactions! Write down quickly whatever comes to mind.

Step 2. Explore your Step 1 reactions by asking, Why did I react the way I did to this question? In other words, think about your reaction—that is, *think about your thinking*—and see what you come up with. Note: this will require much more effort than Step 1.

Step 3. Return to the question, "What does it mean to you to live in a universe?" but this time rather than simply reacting to the question, reflect on it. Give this a full fifteen minutes, firm in the conviction that you have within you a better, deeper, more profound response to this question than what you offered in Step 1. Persevere. Note down what you discover.

Step 4. Not done yet! For a final time go back to the same question—"What does it mean to you to live in a universe?"—but this time, rather than answering from your head—that is, in a pragmatic or rational way—answer the question with your whole being. In other words, allow this question to inhabit your entire body (e.g., your gut, your heart). Perhaps even hold your hand over your heart as you contemplate. The point is to feel the question alive in you—and then simply be attentive to what feelings and thoughts arise in response. Best yet, go out into the dead of night and pose this question to the heavens! When you are ready, reflect (There's that word again!) on your experience in moving from reaction (Step 1) to cerebral reflection (Steps 2 and 3) to *full-bodied* reflection (Step 4). What was this process like for you? Why? What did you learn?

2

COMING TO AWARENESS

A Living Planet

Imagine Earth is your biological mother—because, in a very real sense, she is. Imagine the Sun is your biological father—because, in equally real, life giving ways, he is. Imagine that after the spirit of God touched them, your distant but brilliant father and 70-million-square-mile mother not only fell in love, but began making love: imagine Ocean and Sun in coitus for eternity—because they are. Imagine your ocean mother's wombs are countless, that her fecundity is infinitely varied, and that her endless slow lovemaking with Sun brings about countless gestations and births and an infinity of beings: great blue whales and great white sharks; endless living castles of coral; vast phalanxes of fishes; incalculable flocks of birds; gigantic typhoons; weather patterns the size of continents—because it does.

—David James Duncan[1]

The language, above, in Duncan's passage—"Earth is your biological mother . . . Earth and Sun in coitus . . . endless slow lovemaking"—may sound peculiar because we have grown up in a culture that tends to view Earth as an object, a supply house, a substrate that hosts human activities. But as Duncan reminds us, Earth is not an object any more than you and I are objects. Earth is an infinitely complex, nurturing womb for life. As our liver cells or skin cells or muscle cells are to us, so *we are* to Earth. Yes, Earth is our larger body. Earth does not belong to us; rather it is we who belong to Earth; or as theologian Thomas Berry liked to say: "The human is derivative; Earth is primary."[2]

What about you, dear reader—what is you experience of Earth? Do you *know* Earth through mud on your feet, wind ruffling your hair, pollen dust on your table, the wren nesting by your bedroom window? Or do you think of Earth as a NASA photo showing a blue orb with a mantle of clouds? In other words, is your understanding and experience of Earth from the outside looking down, akin to looking at a two-dimensional picture, or do you know and experience Earth from the inside—in an embodied, direct, subjective, visceral way?

Your answer matters insofar as distancing from, and objectification of, Earth cripples our consciousness. And this really has to stop. After all, Earth is our *spaceship*; it's all we've got. And so it is high time that we develop the capacity to see Earth with new eyes—the eyes of relatedness. In this chapter, I offer stepping stones toward this objective.

FOUNDATION 2.1: EARTH—THE *MOTHER* OF ALL

What is Earth to you? A rock in space? A planet in the solar system? A bundle of resources? Your home? Your mother? Your larger body? The answer you give is a reflection of your "story" and, as such, it affects, mightily, your mode of consciousness—that is, how you perceive reality. To explain what I mean I ask that you take a walk with me now. Rather than a pretend walk, make it the real thing. Leave your domesticated life behind and step into the wild world of wind, water, and dirt. As you step outside, remove your shoes—those fabricated contrivances that separate your flesh from the *flesh* of Earth.

With each step, I invite you to pay attention to how the living Earth pushes up, probing the contours of your feet and how your feet press against and, in a sense, massage the ground. Allow your extraordinary feet, those "down-turned hands at the end of your hind legs," to direct your awareness.[3] Feel the coolness of Earth, the textures underfoot, the sounds, both soft and crackly, created by each footfall. Notice, too, how this simple presencing of the world underfoot leads you to slow down and pay attention to the movements, sounds, and smells around you. You are leaving "chronos" and entering "kairos," the present-moment experience of time (see chapter 1). As you slow, notice that there is more going on here than meets your eyes. Consider that just as your feet reach out and touch Earth's *skin*, there are tens of billions of tiny organisms living on and in the soil where you are treading and that these beings are sensing you. Yes, wherever you walk, the beings underfoot *know*, in their particular way, that you are present here.

Consider, too, how, with each step, the Earth is there to receive you, hold you; there is nothing you need to do. It is, of course, the amazing force of gravity that is securing you to Earth. So, experience gravity for the wonder that it is. See it as the attraction between two bodies—two subjects—the magnetism between two lovers, the primordial attraction of the body of Earth for the body that is you. This bonding of bodies—of matter—is a fundamental organizing principle of your home, planet Earth.

As you continue to walk along, give yourself permission to become even more deeply delighted by the forgotten pleasures of earthly contact. In this moment you are no longer pacing indoors on top of a suspended concrete slab. No, now your flesh is in touch with your larger body. You are reclaiming part of your birthright. After all, the word "human" has its origins in *humus* or soil; just as the word "matter" has its roots in *mater*, or mother. With your bare feet, you are now in direct contact with the "mater" that nurtures and sustains you. All along, you may have thought it was Wal-Mart that supplied your needs, but now, possessed of new planetary eyes, you realize that Wal-Mart is just a small-bit subsidiary of planet Earth.

Now, the time has come to find a seat on the ground, not on a chair, mind you, but on the dirt that is Earth. By sitting directly on Earth you are reenacting what your human ancestors did for thousands of generations. Indeed, sitting in chairs and sleeping suspended above the ground is a very new practice for humans—one that, while offering certain comforts, separates us from direct contact with Earth. So, with the innocence and inquisitiveness of a child, explore what it is like to have your spine in direct contact with Earth.[4]

As you settle in, behold all the green that surrounds you. You are looking at Earth's most fundamental adornment. Leaves, whether of herb or grass or tree or algal cell, are the foundation of Earth's fecundity on both land and sea. Without leaves there would be no photosynthesis, no me, no you!

And now notice how you are being breathed. You don't do it consciously; your body breathes you—twelve inhalations each minute, seventeen thousand each day. You live in a veritable sea of air; the air sustains you. As you quiet yourself further, begin to experience this air with new sensibilities: "not as a random bunch of gases simply drawn to Earth by . . . gravity, but an elixir generated by the soils, the oceans, and the numberless organisms that inhabit this world, each creature exchanging certain ingredients for others as it inhales and exhales, drinking the sunlight with our leaves or filtering the water with our gills, all of us contributing to the composition of this phantasmagoric brew, circulating it steadily between us and nourishing ourselves on its magic, generating ourselves from its substance."[5]

Finally, take note as a mosquito circles around your head, bobs up and down, and then alights on the back of your left hand.[6] This mosquito hasn't arrived randomly; it has found you by following your chemical emanations. Yes, everything has its particular musk; you are part of the olfactory field of this mosquito. Your first impulse is to squash it, but this time your curiosity overrides your annoyance. You lift your hand up close to your face to watch as the mosquito hunkers down and then pierces your skin with its needle-like proboscis. With macabre fascination you continue to observe as the mosquito's abdomen fills with your blood. Again the impulse to squash this freeloader arises but, once more, you hesitate. By extending attention to this mosquito you are able to see, perhaps for the first time, the similarities between it and you. For just as you take delight in savoring the world's scents and tastes, you realize that blood also has its own mix of sweet and sour, salty and savory. You even manage to smile a little bit as you realize that you are participating in nature's great gift economy—offering some of your life force to this creature. It seems like a pitifully small offering for all that you receive from Earth, day by day, but an offering nonetheless. As David Abram reminds us in his splendid book, *Becoming Animal*, "No matter how earnestly we humans strive to exempt ourselves . . . we cannot escape our participation in [nature's] cycles of exchange. If we ingest the land's nourishment . . . incorporating the world's flesh into our own—it can only be so because we, too, are edible. Because we, too, are food."[7]

After the mosquito departs, you look down at your hands and are amazed by them. Heretofore, you have simply taken your hands for granted, but now, unexpectedly, it is as if you are seeing them for the first time! And you find yourself asking: What are these hands? Where did they come from? Why are they shaped as they are? With childlike curiosity you explore the texture and suppleness of your hands, the lines on your palms, the roundness of your knuckles, the hardness of your fingernails, the mechanics of each hand's movements. You consider, as an anthropologist might, the paw nature of your hands—your grasping fingers and opposable thumbs, the webbing at the base of each finger. You note that your hand is not encased with a heavy shell or a thick pelt; it lacks fearsome claws. Rather, it is soft, vulnerable, designed for doing, sensing, experiencing, knowing.[8]

The form of your hand has been shaped by Earth. Only a planet with Earth's particular characteristics could produce the particularity of the human hand, and it has taken five billion years of conditions unique to

Earth to hone the human hand into its present form. Indeed, your hands are the result of a long evolutionary process. Prototypical hands first appeared long ago in the form of primitive fish fins in the primordial seas where life first arose. Your fish ancestors learned to push up onto dry land, and eventually—once incarnated as mammals—learned to use their limbs to climb and, with the addition of the opposable thumb, to grasp onto tree limbs. With the passage of more time, early primates descended from the trees and learned to use their remarkable hands to make fire, fashion tools, and plant seeds.

The particular conditions offered by Earth have shaped everything about you. Not just your hands but your eyes, with their capacity to distinguish shape, distance, color, depth, and shadow; not to mention your ears, with their ability to detect the wide array of sounds that ensure both your survival and your pleasure, and the same is true for every other feature of your biology, as Abram summarizes:

> This ageless intercourse between [your] body and . . . [the body of] earth—this co-evolution—has shaped the organs and tissues of every earthly organism. . . . [So it is that] we humans are corporeally related, by direct and indirect webs of evolutionary affiliation, to every other organism that we encounter. Our nervous systems . . . are wholly informed by the particular gravity of this sphere, by the way the sun's light filters down through Earth's atmosphere, and by the cyclical tug of Earth's moon. In a thoroughly palpable sense, we are born of this planet, our attentive bodies coevolved in rich and intimate rapport with the other bodily forms—animals, plants, mountains, rivers—that compose the shifting flesh of this breathing world.[9]

To ignore our profound embeddedness in Earth as inconsequential diminishes our consciousness, severely circumscribing our experience of what it means to be human.

Exchange Is the Essence of Life on Earth

If I were asked to give a single-word descriptor of life, I would choose the word *exchange*. Whether human or frog or worm or bacterium, the survival and flourishing of all that lives depends on constant exchange with both other organisms and the physical environment. Indeed, all creatures, no matter how small, are literally *designed* for the exchange of the minerals, gases, nutrients, and water that all life needs for survival. For example, it is on the surfaces of your lungs that Earth's air first touches and then enters your body; likewise, it is on the surfaces of your intestines that Earth's food enters your blood stream. Insofar as exchange surfaces are a basic design feature for all life, it should come as no surprise that much of the living world is composed of these surfaces (figure 2.1).[10]

It is natural to suppose that most of planet Earth's exchanges would occur on the leaf surfaces of land plants, since plant leaves are highly visible and abundant, from a human perspective. And, in fact, if all the leaves of Earth's land plants were stitched together, they would make a complete covering for the Earth—that is, the cumulative surface area of all plant leaves is roughly the same as Earth's total surface area.

And what about the seas that cover roughly 70 percent of Earth? It turns out that the combined surface area of the tiny planktonic organisms that dominate the world's oceans is approximately six times greater than Earth's total surface area. This seems odd insofar as the actual mass of oceanic phytoplankton is tiny compared to the mass of leaves on land. But appearances can be deceiving. The paradox is resolved by the fact that—unlike land plants—most sea plants don't possess multicellular leaves but, instead, occur as tiny, free-living individual cells. This makes a huge difference when it comes to surface area. For example, imagine dispersing the hundreds of

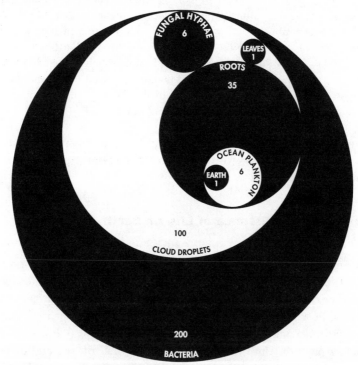

Figure 2.1. The surface area of Earth relative to other life surfaces.

thousands of cells in a single tree leaf, one by one, into the ocean; to do so would magnify the total surface area of that single tree leaf dramatically, which is precisely the strategy of ocean plankton—that is, to be small and numerous. If this still seems obtuse, imagine that I gave you the task of measuring the exterior surface area of a medium-sized log. Then, suppose that, after completing your task, I ran that log through a wood chipper and asked you to recalculate the surface area of the log. Before you even began you would know that the cumulative surface area of all those chips (think individual tree-leaf cells) would be many times greater than that of the external surface of the original log.

But the surface area of the foregoing categories pales in comparison to the surfaces of the most abundant of all organisms, bacteria. Lest you imagine that bacteria are insignificant creatures, think again: Bacterial surfaces exceed Earth's surface area by some two-hundred-fold. Again, bacteria achieve their enormous cumulative surface area by being small and very numerous. Given the importance of surfaces for life's exchanges, it should come as no surprise that the air, water, and soil, in which all life is embedded, are also rich in surfaces. The surface area of Earth's topsoil is estimated to be more than ten thousand times the surface area of Earth itself; and any way you cut air, it's all surface; ditto for water.[11]

SOLID MATTER OR EMPTY SPACE?

I am struck by how difficult it is to see life's myriad surfaces. The bacterial, fungal, and algal exchange surfaces, where most of life's exchanges take place, are invisible to me. Meanwhile, the things that I do see are, in a sense, illusions. For example, when I look in the mirror, I see a reflection that is the result of light bouncing off the atoms that make up my skin tissue. Although I look solid, I know that I am mostly emptiness—visible emptiness! Cosmologist Brian Swimme explains it this way: "You are more fecund emptiness than you are created particles. We can see this by examining one of your atoms. If you take a single atom and make it as large as Yankee stadium, it would consist almost entirely of empty space. The center of the atom, the nucleus, would be smaller than a baseball sitting out in center field. The outer parts of the atom would be tiny gnats [electrons] buzzing about at an altitude higher than any pop fly Babe Ruth ever hit. And between the baseball and the gnats? Nothingness. All empty. You are more emptiness than anything else."[12] The reason that the chair you sit in or the cup you drink from seems to be solid is that electromagnetic forces operate to hold the speck-like nuclei of atoms together. The more I learn about the world, the more I am able to accept, and even appreciate, that things are not as they appear to be.

In sum, life probes and plants itself into the surface-rich matrices formed by soil and air and water. "These matrices bathe life, harbor life, surround life, nourish life."[13] Seeing the locus of life's action as surfaces helps us remember that the essence of life is exchange—giving and taking—in short, relationship.

Earth's Metabolism

All that lives must eat. Even plants are eaters—that is, they feed on solar photons with their leaves while they slurp up water and nutrients with their roots. All that walks, crawls, wriggles, runs, flies, swims, hops, and burrows relies, either directly or indirectly, on the food-manufacturing skills of plants for survival. Chances are that ever since grade school you have heard, repeatedly, how plants use energy from the sun to create complex sugars from carbon dioxide and water. Perhaps because you have heard it so often, the wonder of it escapes you. But consider, with fresh eyes, the magnificence of photosynthesis: Two formless things, water and air, forged into a solid: a molecule of sugar. Look around at the vegetation close at hand—the wildflower meadows, cattail marshes, pine forests—and remind yourself that all this plant life was sparked into being through photosynthesis—that it is the *miracle* of photosynthesis that shapes the formless into form.[14]

The star performer in photosynthesis is the 120-atom chlorophyll molecule. Chlorophyll is designed to snag very tiny and very fast-moving particles of light known as "photons." Normally, when photons strike a surface, they are converted into heat; but chlorophyll in leaves grabs photons in just the right way so that their energy can be used to initiate photosynthesis. Fortunately, the world is awash in chlorophyll. How much? Here's a way to think about it:

> Imagine the human population (seven-plus billion) increased by a few ten-million-fold. This multitude, reduced to chlorophyll molecules, would comfortably find niches latched onto internal cellular membranes within a single square centimeter of leaf area. That's the same scale as the density of photons striking the same leaf surface. It enables each chlorophyll molecule to absorb a couple of photons per second from a shower of bright sunlight.[15]

In spite of chlorophyll's effectiveness at capturing light, less than 1 percent of the solar radiation striking Earth's surface is used in photosynthesis. The rest of the Sun's energy also goes to good use. For example, a significant fraction is used to heat Earth's air and land and water. This heating of Earth's surfaces creates winds that, in turn, power Earth's weather systems. Another major chunk of the Sun's energy is used in the evaporation of moisture from water, soil, and leaf surfaces, thereby activating the water cycle. In sum, all of our weather—wind, clouds, rain and snow, hot and cold fronts, all of it—is driven by energy from the Sun.

Life in a Corn Field

Earth's millions of distinct species are able to coexist, in large part, because they feed on plants in specialized ways. For example, imagine that you are in the middle of a corn field. It is the end of the growing season and, as you look around, you see mature corn plants everywhere. Examining these plants, you note that certain insect species are chewing holes in the corn leaves while other insect species are boring their way into the corn stalks. Curious, you yank some corn plants from the soil and inspect the roots with the aid of a small portable microscope. There you discover a multitude of protozoans, nematodes, and insect larvae feasting on the corn roots. Taken as a whole, these plant eaters, or herbivores, feeding on the corn plant's leaves, stalks, and roots might take in 10–20 percent of the corn's total production. Meanwhile, humans harvest another 10–20 percent of the corn plant's production as corn kernels. So, all told, roughly one-third of a living corn plant is eaten fresh by other organisms.

And what actually happens to all this corn tissue consumed by herbivores? Some of it is added to the eater's mass as new growth, but, perhaps, not to the degree that you might expect. In fact, on average, only about 10 percent of the plant material consumed by herbivores results in new growth. And the rest? Well, when we humans act as herbivores and eat leaves (lettuce), or flowers (broccoli), or grains (corn kernels), about a quarter of our intake passes through our gut undigested; most of the rest is "burned" in respiration to maintain our normal body functions (e.g., breathing, moving, thinking). At most, only 10 percent goes to growth (i.e., weight gain or tissue replacement). The same is true for the blackbird that eats the berry or the red-tailed hawk that feeds on the blackbird. Although the percentages may differ somewhat depending on the type of food and the type of organism doing the feeding, 10 percent is a reasonable estimate for the overall efficiency of energy transfer in feeding relationships among organisms in nature. In fact, ecologists refer to it as "the 10 percent rule."

This rule reveals why it is more ecologically efficient for humans to sustain themselves by eating plant foods, such as grains and beans, than to first feed these foods to animals and then to eat the animal flesh. Eating plants directly allows us to incorporate 10 percent of the plants' energy for

A WORLD WITHOUT DECOMPOSERS

Without decomposers we would be in a world of trouble. To make this point more explicit, consider how a forest works. Each year trees grow larger and gradually, with the passage of time, old trees die and fall to the ground. Tons of dead tree leaves also drop to the forest floor each year. Now, imagine that there were no decomposers and that all this dead plant stuff simply accumulated on the floor of the forest. If this were to happen, after a few hundred years there would be more than ten feet of dead tree debris on the ground, making it very difficult to even enter the forest. Worse still, if there were no creatures around to break down the dead leaves and the downed trunks and, thereby, release nutrients back to the soil, the entire forest would eventually weaken and die for lack of nutrients. But, of course, this is not the way things work. All the dead organic matter, in the case of the corn field and the forest, doesn't accumulate; instead, it provides food for trillions of decomposer organisms—such as beetles, worms, slugs, mites, and nematodes, not to mention bacteria and fungi—that are largely hidden from sight.

our body function and growth; whereas feeding those plants to animals (10 percent efficiency), then eating the animals (10 percent efficiency again), only allows us to capture about 1 percent of the plant's original energy.

Returning to the corn field: Given that only one-third of a corn field's production goes to herbivores, including humans, what about all the uneaten roots, stems, and leaves of corn plants lying around after the harvest? As you have probably observed for yourself, eventually this material dies and settles to the ground (barring removal for silage). You might think of it as "waste" but, actually, it provides an energy source for minute, inconspicuous creatures known as "decomposers." It is decomposers that do the critical work of breaking down dead organic matter, ensuring that the nitrogen and phosphorus and other plant nutrients in plant remains return to the soil where they become available, once again, for uptake by plant roots.

One need only go out to a patch of forest with a magnifying glass and get down on hands and knees and paw around a bit in the soil to discover a vibrant "city" teeming with diverse "citizens":

These citizens are often in motion, hurrying along the vast expressways made by moles, the boulevards of earthworms, the alleys between particles of sand or clay, the dank canals that these alleys often become. Certain districts and certain intersections—mainly close to the roots of plants—get especially busy. The citizens move in the dark, sniffing out chemical trails. They are constantly doing business with one another. They traffic in molecules: minerals, organic compounds, packets of energy.[16]

In the early stages of decomposition, ants, worms, millipedes, beetles, and their brethren play an important role by physically *chewing* apart the plant debris, greatly increasing the surface area available for later action by bacteria and fungi. Finally, only tough—difficult to digest—plant materials, like lignin, are left; and these recalcitrant compounds become dinner for certain specialized types of fungi.[17]

Life as Lifting Up and Flowing Down

In the big picture, photosynthesis is the "lifting up" process of life whereby solar energy is used to build new tissue. Decomposition, by contrast, is life's "flowing down" process, whereby dead tissue is broken down, releasing energy to decomposers and nutrients to plant roots.[18] Indeed, everything in the natural world can be thought of in terms of a lifting up or a flowing down process. The vigorously growing garden plants of summer are lifting up, while the leaves decaying on the ground in fall are flowing

---

RECYCLING: FLOWING DOWN IN ACTION

As a beginning student of ecology, it occurred to me that when it comes to recycling, forests have a lot to teach us. But I didn't have the opportunity to fully appreciate the amazing efficiency of recycling until I confronted a seeming paradox in the tropical forest of the Rio Negro region of southern Venezuela. The soils in this region are very old and extremely infertile. Yet the forest, for the most part, is lush and well developed. This paradox—a productive forest ecosystem supported by nutrient-impoverished soil—began to resolve itself the instant I entered the forest. Yes, there were strange sounds and riotous vegetation and the air was thick with moisture. I expected this, but I didn't expect to be literally walking on a sponge of fine roots. Indeed, roots were actually growing on top of the soil in a dense mat, sometimes a foot or more thick. These fine surface roots grew over and around the dead leaves and the woody debris that littered the ground. At one point I bent down to examine a dead branch only to discover that the "branch," below its external veneer of bark, was actually a mass of roots. What was, at first, a paradox now made sense. These Rio Negro forests were able to flourish on infertile soils because their roots grew right up out of the soil garnering the precious nutrients in the fallen leaves and branches before they could leach away.

down. The cow pie in the pasture flows down, releasing nutrients to the soil that are lifted up in the form of new grass growth.

These paired concepts of lifting up (plant production) and flowing down (decomposition) allow us to see the natural world in a fuller, more dynamic way. For example, imagine that while walking along a stream-bottom trail, you come upon the carcass of a female deer. You crouch down to have a closer look. Judging from her size, you conjecture that this deer died in her second year of life. During her life, her body was fueled by acorns, berries, fresh plant shoots, and, perhaps, corn from nearby fields—all products of photosynthesis. But now, in death, you note that myriad organisms are finding sustenance in her body.

Educator Paul Krafel has described the remarkable parade of organisms that feast on a deer carcass. First, the vultures arrive and then the blowflies swoop in to lay their eggs. The tiny larvae emerging from the blowfly eggs burrow into the deer's flesh to feed, and a portion of the dead deer is thus transformed into writhing masses of maggots. Then, predators such as the rove beetle visit the deer to feed on the maggots. At the same time, there are the robber flies hanging out around the perimeter, ready to catch flying insects attracted to the carcass. Looking closely, one might also discover parasitic wasps injecting their eggs into insects feeding on the carcass. In the final stages of this feast, new varieties of insects come to feed on the doe's greasy bones and fur. All told, the energy within the deer carcass nourishes more than a hundred different species of animals. Most of the deer flows down, rendered into gas, water, and mineral elements. Meanwhile, some 10 percent of the deer (in accord with the 10 percent rule) is momentarily lifted up, contributing to the growth of new tissue in creatures such as vultures, maggots, and beetles.[19]

And, of course, it is no different with you and me: Simply by eating food (i.e., parts of plants and animals), we are involved in Earth's lifting up dynamic, and when we excrete wastes, we participate in the flowing down process. Our big *flow down* will occur when we die. When that happens, your body, like mine, will probably be slowly devoured by microbes, rather than predators . . . but one never knows! Indeed, there was a time not long ago when I came face to face with a panther while walking along a forest trail at midday in northern Brazil. I was lost in thought when, suddenly, the panther appeared just thirty yards in front of me. I stopped; the panther stopped. We gazed at each other for perhaps a minute. I was dead meat; there was no escape for me. Luckily, there was nothing menacing in that panther's posture; he/she seemed more curious than hungry and quietly slipped away into the undergrowth.

FOUNDATION 2.2: EARTH'S CYCLES

One of my hopes is that you will finish this chapter with a heightened capacity to see the myriad cycles operating in and around us. Indeed, you are participating in a planetary cycle in this very moment simply by breathing. That's right, in the time it takes you to read this sentence, millions of molecules of oxygen from Earth's atmosphere will have entered your bloodstream and on your next out breath millions more molecules of carbon dioxide—the byproduct of cellular respiration—will leave your body and enter the atmosphere. In this vein, it is worth noting that the atoms that presently make up your body will only stay with you for a short time—weeks, months, a few years, at most—before leaving and being replaced by new atoms from outside of you. Everything is cycling in and out all the time. So it is that with each inhalation, you take into your body oxygen atoms that were once part of worm, tree, lake, and rock; and with each exhalation you give back part of what was once you to Earth.

Earth's Cycles Are Essential for Life's Flourishing

Observe rain falling and you are seeing, up close, Earth's water cycle in action. This cycle is particularly important to life because most organisms—humans included—are more than two-thirds water. Indeed, "water is the raw material of creation, the source of life. . . . Life originated in the oceans and the salty taste of our blood reminds us of our marine evolutionary birth."[20]

In the hydrologic cycle, water vapor rises from the surfaces of lakes, oceans, plants, and soil through the Sun-driven lifting-up process of evaporation and then, after a time, falls back down to Earth as precipitation. The precipitation falling on land each year seeps into the soil where it may be quickly lifted up by tree roots. Alternatively, this precipitation might find its way into nearby rivers and thence to the ocean, or it might slowly percolate down through the latticework of soil particles and eventually make its way to aquifers—i.e., water reservoirs deep underground.

And let's not forget our part. Each day water—approximately two quarts of it—finds its way into each of our bodies, either through the liquids that we drink or as a constituent of the food we eat. This water nurtures us insofar as it is only in the presence of water that the cells of our bodies are able to absorb nutrients and secrete wastes.

All of life's important elements have cycles, not just water. Sometimes, as with nitrogen, the cycles have a gaseous phase. In other cases (e.g., calcium, magnesium, and phosphorus), the cycles are limited to a solid and liquid

OUR PROFOUND CONNECTION TO THE WATER CYCLE

I don't think I fully grasped my watery nature until I was lying in bed one evening during a spring rainstorm. As is often the case, I experienced an abiding sense of peace listening to the rain settle onto the land. Why is it, I wondered, that I find it so comforting to be in the presence of rain? And then, suddenly, I knew. It's because rain—water—is what we humans mostly are! It is the medium that brought us into being. In the most intimate of human acts, spermatozoa from our father were set free to literally swim up a canal in our mother toward an egg. The sperm joined with that egg and we began to grow. And, then, for the first nine months of our lives we lived in a salty sea of amniotic fluid.[21] And still now, as adults, our bodies are mostly water. Yes, we carry around some forty quarts of salty water in our trillions of cells.

phases. When a cycling element serves as a building block for life, the term "nutrient cycle" is used.

One way to grasp the remarkable journeys of life's nutrients as they cycle is to follow the meanderings of a single atom through time. For example, imagine for a moment that you are (literally!) a calcium atom that has just been weathered free from a rock buried in the soil of a young forest. Initially, you find yourself dissolved in a drop of water lodged between two grains of sand, but soon you are sucked up by a tree root and quickly incorporated into a leaf. Then, as winter sets in, the leaf you are encased in is shed to the ground. Come spring, you are absorbed by a bacterial cell feeding on that decomposing leaf. Then an earthworm, ingesting the remains of the leaf, takes you in only to deposit you, a day later, as a casting. In no time, you again find yourself being lifted up by a tree root. This cycle repeats many times—soil to tree and back to soil—but a time comes when you find yourself encased in an acorn, which is carried off by a squirrel. On a frigid winter afternoon, the squirrel consumes the acorn. This same squirrel later dies by the edge of a stream and is washed into the water during a spring flood. Now, as you float down the stream—a calcium atom entombed in the dead squirrel's flesh—microbes set to work on the squirrel carcass and you are set free—but not for long. An algal cell absorbs you into its tiny body; and then the larva of an insect eats the algal cell; and then that wriggling larva is gobbled up by a small trout. When the trout dies, you are again set free by decomposers that dismantle the trout's body.

Decade by decade, you join one organism after another. Meanwhile, the inexorable force of gravity is taking you further downstream until one day you reach the ocean. There you continue to be shunted back and forth from organism to water until one day you find yourself lodged in the bone of a sperm whale. When the whale dies, its bones settle to the ocean's depths and, over time, become buried in sediment. At this moment, it might seem that your journey as a calcium atom is over. It is certainly possible that you may remain locked in ocean sediments for millions or even billions of years. However, a time may eventually arrive when the clashing of Earth's tectonic plates will thrust those sedimentary deposits (of which *you* are a part) upward in a fit of mountain building. Then, chemical and physical weathering processes will be able to act on the sedimentary rock that encases you, inviting you once again back into the cycle of life.

A Nutrient Cycling Case Study

Scientists have spent a lot of time attempting to understand nutrient cycles insofar as these cycles are essential for a healthy planet. They conduct their studies by charting the movements of life-giving nutrients, such as nitrogen, phosphorus, and calcium. This all sounds somewhat abstract, so let's look at a real-life example. Some years back, two American ecologists, Herbert Bormann and Gene Likens, initiated an ambitious nutrient-cycling study in the mountains of New Hampshire. Their aim was to learn about the movement of nutrients in a forested mountain watershed. The stream running through this watershed—Hubbard Brook—collects all the water draining from the surrounding mountains.

Bormann and Likens conceptualized the watershed as a network of boxes or compartments—for example, trees as one compartment, soil another, and so forth—each "box" containing a stock of nutrients. The nutrients in any given ecosystem box might move back and forth into other ecosystem compartments as shown by the arrows in figure 2.2.[22]

The starting point for their work was to determine the total amount of each nutrient in the various ecosystem boxes. Imagine the challenge: You walk out into the watershed to determine the amount of, say, calcium in just the forest-tree compartment. How would you do it? In a way it's just common sense. First, you would have to know the total weight (mass) of the forest and then you'd have to figure out the percentage of that mass comprised of calcium. But weighing the whole forest, even if it could be done, would leave you with a huge pile of logs, branches, leaves, and roots; and, of course, you would end up destroying the very forest you had hoped to study. The solution, rather than cutting down and weighing each tree in the watershed, was, instead, to measure the diameter and height of all the living trees and estimate their

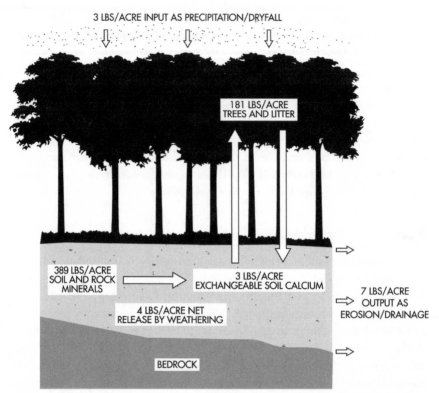

3 LBS/ACRE INPUT AS PRECIPITATION/DRYFALL

181 LBS/ACRE
TREES AND LITTER

389 LBS/ACRE
SOIL AND ROCK
MINERALS

3 LBS/ACRE
EXCHANGEABLE SOIL CALCIUM

7 LBS/ACRE
OUTPUT AS
EROSION/DRAINAGE

4 LBS/ACRE NET
RELEASE BY WEATHERING

BEDROCK

Figure 2.2. Schematic map, Hubbard Brook ecosystem, showing stocks of calcium in various forest compartments and calcium fluxes between compartments and into and out of the ecosystem.

weight based on those measurements. This approach yielded a good estimate of the weight of the entire forest compartment. Next, the researchers collected samples of all the tree-tissue types—leaves, bark, trunks, roots, and so on—and used prescribed chemical testing procedures to determine the concentrations of calcium and other important nutrients in each tissue type. Then, by scaling up, they were able to estimate the total amount of each nutrient element in the entire aboveground forest vegetation compartment. Similar approaches were taken to estimate the nutrient stocks in other ecosystem compartments (e.g., roots, soils, animals).

If you think this kind of work is tedious, I can affirm, from my own experience, that it is! And there is more. The researchers also had to determine the quantity of each nutrient entering (inputs) and leaving (outputs) the watershed. Nutrients entered the watershed in rain and snow and as dust particulates. Precipitation collectors were used to measure these atmospheric inputs, while a stream-monitoring station at the base of the watershed was used to monitor the quantity of water leaving the watershed. By conducting chemical analyses of both the precipitation coming into the forest and the stream water leaving the ecosystem, the researchers were able to calculate the quantities of nutrients entering and leaving the watershed.

So, what's the punch line, the end result of all this toil? Aside from revealing details on the how and when and why of nutrient movements in forest ecosystems, the study showed that New England forests are very thrifty—that is, they don't waste much. For example, only 7 pounds of calcium per acre exited the Hubbard Brook watershed each year, whereas 3 pounds of calcium entered the watershed per acre via precipitation each year. Hence, the net annual loss of calcium was only about 4 pounds per acre per year, a tiny amount compared to the roughly 570 pounds of calcium per acre in the soil and vegetation compartment (figure 2.2). Thus, the researchers concluded that this forest watershed, in its natural state, was extremely efficient, retaining 99 percent of its calcium from year to year. The little bit of calcium that was lost was presumably replenished by the weathering of rocks in the soil.

It is the same with forests everywhere: Nutrients cycle efficiently so long as humans don't interfere with natural decomposition and plant nutrient-uptake processes. However, when forests are significantly disturbed by such things as careless logging practices, they begin to hemorrhage, leaking their nutrient stores. The reason is that normal nutrient-conserving mechanisms—such as well-developed root systems and thick, organic-soil coverings—are damaged, and when this happens forest vigor and well-being are compromised.

Nutrient cycling isn't restricted to forests; it occurs everywhere, all the time. Indeed, each day each of our individual behaviors can enhance or disrupt Earth's nutrient cycles. For example, putting lawn leaves in plastic bags and sending them to the local landfill (as many Americans do), isolates these organic materials from natural nutrient recycling processes. This need not be. Better to keep household nutrients in circulation by composting leaves and kitchen scraps and using the resultant humus to enrich household gardens.

A Zero-Waste World

The overwhelming majority of what Americans purchase ends up being trashed, not recycled. As a result, each American, on average, throws away roughly four pounds of materials each day; this adds up to about 430 billion pounds of trash buried in landfills and/or burned in incinerators each year in the United States. This stuff that we call "garbage" or "trash" consists, in actuality, of Earth elements—that is, all of it originated from planet Earth. And though we may think we are throwing it "away," there really is no "away."

In the natural world there is no such thing as trash or garbage; there, everything is recycled. The same could be true for us; a zero-waste world is possible. We could start with all the so-called waste that we flush down our toilets. Growing up in America, I was taught to flush my droppings. This seemed to make sense; I never questioned it. But then one day, in my midtwenties, I had to poop and there was no flush toilet anywhere nearby. I was in the rain forest in Amazonia at the time and, though I had had experience with pooping in the woods, this was to be my first tropical squat. Undaunted, I looked around and located a nice spot and relaxed into it. I had barely finished when a large beetle came careening through the forest and landed on my steaming deposit. At first, I wondered what this creature could possibly want with my organic leftovers. It was clear that the beetle hadn't arrived by mistake; she was going to stay; so I decided to stay and watch. She was about the size of a medium-sized pebble, with a hard, shiny body. Slowly she began shaping part of my deposit into a small ball, which she then proceeded to bury in the soil. This creature, I later learned, was a dung beetle and was using my poop as a substrate for her eggs. Eventually, when these eggs hatched, they would be surrounded by "food" contributed by me. So, what I had always regarded as my waste was just what this dung beetle needed to start raising her family.

This revelation prompted me to see how very strange it is that in the United States we take two perfectly good resources—our human manure and fresh water—and splat them together in the toilet bowl, making them both useless. Fortunately, there are now more enlightened technologies available, such as waterless urinals and composting toilets, which, when used, preserve the integrity of each of these *resources*, while avoiding the harmful *downstream* effects created by our present habit of flushing away our organic leftovers.

Of course, it would take a good bit of educating to befriend our waste, so, what better place to start than in our schools? In this vein, when engineer John Todd was asked to design a wastewater treatment facility for an elementary school in Toronto, he used nature as his mentor and began with the principle: Human waste equals food. Todd's facility, dubbed a "living machine," is part art, part function, and part teacher.[23] Located in the central atrium of the school and consisting of seventeen water tanks arranged in a snail spiral, the living machine looks like a water sculpture. The school uses conventional water-based toilets. Once the waste is flushed, it is rapidly digested as it passes through four tanks—two with oxygen, two without, each filled with microorganisms. Next, the water is pumped to the highest of the seventeen clear plastic tanks. Here, the students may watch as the water (containing the decomposed residues of their pee and poop) flows first through tanks filled with algae, then through tanks filled with aquatic-rooted plants, and finally to tanks with animals, including clams, snails, and fish. By the time the water reaches an adjacent pond, it is clean.[24]

Variants on Todd's living machine are now being used to process industrial wastes. For example, in their book *Ecological Design*, Sim Van der Ryn and Stuart Cowan describe how a film studio in Wuxi, China, resolved a difficult pollution problem involving silver-contaminated wastewater. The studio introduced water hyacinths into a series of ponds. These water hyacinths, in effect, mined silver from the wastewater and accumulated it in their roots. Silver concentrations in the hyacinths' roots were up to thirty-five thousand times higher than the concentrations in the surrounding wastewater. The film studio harvested the hyacinths' roots, extracted the valuable silver, and then reused it.[25]

It won't be easy to match nature's zero-waste standard, but with some creativity, ingenuity, and a shift in consciousness, it might just be possible. And there are opportunities for each of us to get involved. For example, have you ever heard of "upcycling"? It means taking something that is destined for the trash and inventing a new use for it. For example, imagine finding a discarded 1-liter beverage bottle and instead of putting it in a recycling bin—that is, "downcycling"—you "upcycle" it by transforming into a splash-guard (fender) for your bicycle. Or suppose that instead of merely recycling used cardboard and aluminum, you upcycled these two components, using them to make an outdoor solar cooking oven?[26]

STEPPING BACK TO SEE THE BIG PICTURE

Tell me the story of the river and the valley and the streams and woodlands and wetlands, of shellfish and finfish. Tell me a story. A story of where we are and how we got here and the characters and roles that we play. Tell me a story, a story that will be my story as well as the story of everyone and everything about me, the story that brings us together in a valley community, a story that brings together the human community with every living being in the valley, a story that brings us together under the arc of the great blue sky in the day and the starry heavens at night.

—Thomas Berry[27]

Recently I heard a story about the planning process that went into the construction of the New England Aquarium in Boston. There was a design competition and the adults decided it would be nice to invite some children to help in the process. When the decision on the final design was close to being finalized, someone remembered to ask the kids what they thought. There was a silence and then a little girl raised her hand and asked, "When do we get to put our hands in the water?" Yes, kids, with their *beginner's mind* know that nature is not to look at, but to be with and to dwell within.

Richard Louv in his book *The Last Child Left in the Woods* notes that in recent decades the wild world of streams, forests, and fields has become increasingly off-limits for many children. This is, in part, the result of growing fear. For example, from the 1970s to the 1990s the radius around the home where children were permitted to roam shrank by more than 80 percent. Add to this the growing allure of indoor electronic stimuli—for example, children between six and eleven years old now spend an estimated thirty hours a week gazing at screens—TVs, computer monitors, smart phones—and the inevitable result, it would seem, is a weakening bond between kids and the natural world.[28]

The point, not to be overlooked here, is that we are raising our children in an environment that is utterly new in human history. In this vein, eco-psychologist Chellis Glendinning suggests that modern humans are undergoing, unbeknownst to themselves, a profound form of traumatization. The *Diagnostic and Statistical Manual of Mental Disorders* defines traumatization as "an event outside the range of human experience that would be markedly distressing to almost anyone." In Glendinning's view, "The trauma endured by technological people, like ourselves, is the systemic and systematic removal of our lives from the natural world: from the tendrils and earthly textures, from the rhythms of sun and moon, from the spirits of the bears and trees, from the life force itself."[29]

A growing body of research links emotional and physical health in both children and adults to contact with nature. For example, a study of prison inmates in Michigan revealed that those who had cells facing the interior courtyard had 24 percent more illnesses than those whose cells faced the surrounding farmland. The same thing has been observed in hospitals where patients in rooms with a view of the natural world tend to do better than those without access to such a view. Similarly, researchers at Cornell found that the more contact children had with nature—including such simple things as indoor plants and window views of nature—the better was their ability to cope with stress.[30] These and other studies make the case that direct contact with the natural world centers, heals, and relaxes us.

NATURE DEFICIT DISORDER?

The possible calming effects of nature are of particular interest to researchers seeking to understand the reasons for the dramatic increase in ADHD (attention deficit hyperactivity disorder) among children in recent years. Children suffering from ADHD have difficulty concentrating and paying attention. Scientists at the Human-Environmental Research Laboratory at the University of Illinois have found that when children with ADHD are engaged with nature they show a significant reduction in ADHD symptoms—for example, these children were able to maintain their focus on specific tasks better when in natural settings.

It may well be that a significant factor in the epidemic outbreak of ADHD in recent decades is the growing separation between children and nature, compounded by excessive electronic stimulation. To the extent that this is true, Louv suggests that attention deficit disorder might be renamed *nature deficit disorder*—a condition resulting from the cumulative alienation of humans from nature.[31]

Separation from the natural world—passing our time in indoor "boxes"—appears to have the opposite effect.

Make no mistake: It's not too late for all of us—adolescent kids, college kids, adult *kids*—to cultivate a visceral relationship with the wild Earth that has birthed us. Simply by leaving the *Great Indoors* behind and stepping jubilantly into the juicy shapes, sounds, smells, colors, textures, and movements of the natural world, we create the ground conditions for relationship with life. The first step is to let go of the crippling conditioning that would have us believe that entrance into authentic relationship with the exuberant Earth that has birthed us can happen with an *indoors-only* existence. It can't. Native Americans have something important to teach us in this regard. For them, if you are *inside* you are actually *outside* of the real world and when you venture *outdoors* you are actually *inside*—that is, you are dwelling within what is real and alive and true.

WRAP-UP—EXPANDING ECOLOGICAL CONSCIOUSNESS

I only went out for a walk, and finally concluded to stay out till sundown, for going out, I found, was really going in.

—John Muir[32]

When I was growing up, I saw the Earth as an object filled with stuff—roads, houses, stores, factories, yards, trees, animals—a planet inhabited by lots of separate things. Later, I came to see that I was not alone in this misperception. It seems that most of us have inherited, to varying degrees, a fractured image of planet Earth. We received it from our culture, which transmits, in myriad ways, the message that Earth is an object composed of separate entities, with humans standing apart and above everything else. Hence, it is easy to imagine ourselves as separate from life's myriad cycles—separate from the water, air, and soil—easy to think that we are not part of Earth's vital exchange surfaces—that we stand apart from the feeding relationships that interlace Earth's biota.

I experienced this objectification of Earth in a visceral way recently when I saw a welcome sign at the entrance to a national forest. The sign read, "Land of Many Uses," thereby reinforcing the anthropocentric notion that Earth exists for human use. I was half inclined to cross out "uses" and substitute "beings" or even "cycles." Indeed, a sign reading "Land of Many Beings" or "Land of Many Cycles" would have served as an invitation to see that national forest as animate and beckoning relationship instead of a collection of mute objects for my use.

Summing up, this chapter is an invitation to expand our ecological consciousness by awakening to the interconnected—not fragmented—nature of life on Earth. In spite of appearances, life is not organized as discrete packets—the mouse, the petunia, the jellyfish. Neither the tree outside your window nor you, yourself, are restricted to the solid body that you see. Life's boundaries are not sharp, but soft, porous. As each of us comes to understand this and as we engage in practices to reinforce this understanding, we may reach a point, one day, when the words of David James Duncan (quoted at the beginning of this chapter) no longer seem to be obtuse but, instead, pregnant with meaning:

> Imagine Earth is your biological mother—because in a very real sense, she is. Imagine the Sun is your biological father—because in equally real, life-giving ways, he is. Imagine that after the spirit of God touched them, your distant but brilliant father and 70-million-square-mile mother not only fell in love, but began making love: imagine Ocean and Sun in coitus for eternity—because they are.[33]

APPLICATIONS AND PRACTICES: CULTIVATING AWARENESS

> As I look more deeply, I can see that in a former life I was a cloud. And I was a rock. This is not poetry; it is science. This is not a question of belief in reincarnation. This is the history of life on Earth.
>
> —Thich Nhat Hanh[34]

Planet Earth does breathe, and we breathe with it; Earth does have a metabolism, and we are part of that metabolism; Earth's elements do cycle, and we are part of those cycles.

Participating in Earth's Breath with Awareness

The words *breath* and *air* come from the Latin *spiritus* with connections to soul, spirit, liveliness, emotional vigor, and essence. Insofar as breath is our lifeline, mindful breathing can be a powerful way of cultivating awareness.

My favorite breathing practice is called *breathing with a tree*. You may find this exercise odd and reject it. In fact, I remember making a joke about this practice the first time someone invited me to do it. Like many people, I tend to resist things that I am unfamiliar with. Later, though, I gave this practice a try and was surprised to discover power in it. You may, too.

Start by taking hold of a tree branch. Then, focus on one of the leaves and, as you do this, bring your attention to your breath. As you breathe out, carbon (as carbon dioxide) is passing from your body into the leaf that you are focusing on. Fired by solar rays, this carbon—your carbon—is being forged into sugars inside the leaf. As you breathe in, the by-product of the leaf's solar forging, oxygen, is finding its way into your lungs and thence, via your blood, to the cells of your body. This is happening in real time. Breathe in and then out with full consciousness of the exchange taking place. As you engage in this practice, you give yourself the opportunity to develop a bodily sense of your participation in Earth's breathing.[35]

Participating in Earth's Metabolism with Awareness

Each time we eat we participate in Earth's metabolism. It is easy to forget this, but we can engage, if we choose, in practices to bring awareness to the act of eating. Physician Jon Kabat-Zinn uses the following exercise to help his patients (people who have been living under chronic stress) bring mindfulness to the act of eating.

We give everybody three raisins and we eat them one at a time, paying attention to what we are actually doing and experiencing from moment to moment. . . . First we bring our attention to seeing [one] raisin, observing it carefully as if we had never seen one before. We feel its texture between our fingers and notice its colors and surfaces. We are also aware of any thoughts we might be having about raisins or food in general. We note any thoughts and feelings of liking or disliking raisins if they come up while we are looking at [the raisin]. We then smell it for a while and finally, with awareness, we bring it to our lips, being aware of the arm moving the hand to position it correctly and of salivating as the mind and body anticipate eating. The process continues as we take [the raisin] into our mouth and chew it slowly, experiencing the actual taste of one raisin. And when we feel ready to swallow, we watch the impulse to swallow as it comes up, so that even that is experienced consciously. We even imagine, or "sense," that our bodies are now one raisin heavier.[36]

An important aspect of mindful eating, as dramatized in Kabat-Zinn's raisin exercise, is eating slowly and with gratitude. When people develop the habit of truly savoring each bite of food, from start to finish—instead of starting the next bite before the current bite is finished—they discover that they need to eat less to satisfy their bodies' cravings for flavors.[37] In sum, cultivating an awareness of Earth's metabolism means paying attention to our own metabolism—*what* we eat and, most especially, *how* we eat.

Participating in Earth's Cycles with Awareness

Have you thought about what will happen to your body when you die? If the conventional American burial script is followed, you will be embalmed with highly toxic formaldehyde and then encased in a metal casket, set in a concrete vault in a cemetery. This combination of practices would effectively separate your biological remains from Earth's cycles. But this need not be. In America, it is still possible, in many states at least, for you to be buried in a simple pine box, thereby offering your remains as a kind of banquet for soil organisms. In this more ecologically attuned scenario, after your tissues were broken down to their constituent nutrients, you would be *lifted up* by plant roots, reentering the great chain of life.

In his book *Earth Education*, Steve VanMatre described the mutuality of life and death in a poignant story about a man who had been buried for some time in a small Massachusetts town. For some reason, the people in the town had to dig up the man's grave. When they did so, they discovered that the roots of a neighboring apple tree that had been feeding on the nutrients in the man's body had literally taken the man's shape, from head to toe.[38] Here was a man who, in a sense, was reincarnated as tree.

Cremation is another option that can reunite us, as well as those who mourn our passing, to Earth's elemental cycles. When my father died, our family took the elements remaining from his body to the edge of a small pond that was teeming with life. We lifted his ashes up to the sky and then let them rain down upon the water. I remember being surprised that they sparkled in the sunlight as they cascaded down. At the end, noting a blueberry bush close by, we spread some ashes (an offering of fertilizer) around its base and, then, we each ate a blueberry, knowing that we all participate in the cycle of life and death each moment of our lives.

QUESTIONS FOR REFLECTION

- The point is made in this chapter that, appearances aside, Earth is actually our larger body. Can you offer three personal examples to illustrate how this is literally true?
- Consider that you are mostly empty space: How does this change your notion of your material existence?
- In what sense are you engaged in the biological processes of *lifting up* and *flowing down* each day? And what might you do to bring fuller consciousness to your participation in these processes?
- In what ways might you have lost your essential wildness and become domesticated? What do you see as the possible costs of such loss?
- What behaviors do you engage in that rupture Earth's nutrient cycles and what might you do to avoid or, at least, modify those behaviors?
- If you engaged in the "breathing with a tree" practice, what did you experience and learn in the process? If you were reluctant to try this practice, why do you suppose that is?

❸

CULTIVATING COMMUNITY

Intimacy with Earth's Web of Life

How can we be so poor as to define ourselves as an ego tied in a sack of skin? . . . We are the relationships we share, we are that process of relating, we are, whether we like it or not, permeable—physically, emotionally, spiritually, experientially—to our surroundings. I am the bluebirds and nuthatches that nest here each spring, and they, too, are me. Not metaphorically, but in all physical truth. I am no more than the bond between us. I am only so beautiful as the character of my relationships, only so rich as I enrich those around me, only so alive as I enliven those I greet.

—Derrick Jensen[1]

Driving along a desolate stretch of road in the Everglades in the early 1970s, I encountered a young couple hitchhiking. I picked them up and, seeing their bedraggled appearance, asked them where they had spent the night.

The guy, Dave, responded, "We crashed back in the 'glades. Ya know, trying to get back to nature."

"Whoa!" I exclaimed, "Must-a-been a lot of mosquitoes? How much 'Off' did you use?"

Dave's friend, Sally, said softly, "We don't use repellents."

I probed a bit more, suggesting, "Y'all must have had your blood sucked dry?"

There was a long silence, and then Dave said with a sigh, "Yeah, we got bitten up pretty good but, ya' know, it's all related." Then, Sally turned to me, fixed me with her gaze and added, "When you really think about it, everything is connected—it's all connected."

More than forty years later, I still remember this brief encounter as if it were yesterday. I knew those two were saying something important, but I couldn't quite grasp it. In retrospect, I flag this event as one of my early encounters with *web* thinking.

Now, as a practicing ecologist, I find that wherever I look—whether in Earth's seas or forests or grasslands or streams—I see webs of connections manifesting as communities. Indeed, it is in the nature of life to organize itself into communities. And it is no different with our very bodies. As biologist Bruce Lipton points out: "You may consider yourself an individual, but as a cell biologist I can tell you that you are, in truth, a cooperative community. . . . Almost all of the cells that make up your body are amoeba-like, individual organisms that have evolved a cooperative strategy for their mutual survival."[2] Lipton's point here is that our bodies—by their very biological design and evolutionary history—can best be understood as communities. Genuine communities work together, consciously or otherwise, to serve the whole.

Deepak Chopra, author of *The Book of Secrets*, describes how the community of cells making up a human body exhibits core qualities that are also essential for humankind, as a whole, to flourish. Among these qualities are:

Acceptance: Cells "recognize" each other as equally important. Going it alone is not an option for the community of cells comprising your body.

Awareness: Cells exercise awareness by flexibly responding to surrounding conditions from moment to moment. Getting caught up in rigid habits is not a viable option for the community of cells comprising your body.

Communion: Cells keep in touch with each other—for example, messenger molecules race hither and yon to notify the body's farthest outposts of status and need. Withdrawing or refusing to communicate is not an option for the community of cells comprising your body.

Giving: The primary activity of each cell is giving. Total commitment to giving makes receiving automatic—it is the other half of a natural cycle. Hoarding is not an option for the community of cells comprising your body.

Higher Purpose: Every cell, in effect, "agrees" to work for the welfare of the whole. Selfishness is not an option for the community of cells comprising your body.[3]

Our cells, it seems, live by these five shared agreements—Acceptance, Awareness, Communion, Giving, Higher Purpose—that are very much akin to guidelines laid down in the world's great spiritual teachings. Hence, our cells offer us a concrete model for how we might, as individuals, live respectfully with each other in community.

We know that life goes amok when cells violate their internal code of conduct. Cancer—the unrestrained growth of just one type of cell—is an expression of greed at the cellular level and the consequences are disastrous.[4]

Truly understanding that *everything is connected*—not just grasping it intellectually, but feeling it in our body and bones—engenders intimacy with Earth and, as such, contributes to ecological consciousness. My intent in this final chapter of part I is to invite you even deeper into life's web—so deep that you come to see not just Earth, but yourself, with new eyes.

FOUNDATION 3.1: EACH SPECIES IS A STRAND IN EARTH'S WEB OF LIFE

As a beginning graduate student in ecology at Michigan State, I took a field course with Dr. Patricia Werner. On the first day of class, Dr. Werner took us out to a meadow and invited us to just look around. Then she asked, "What's going on here?" We waited for her to elaborate. She did not. I no longer remember my response, but I do remember Dr. Werner's simple question, "What's going on here?"

This question can be a catalyst for new discoveries. For example, when water condenses on the inside of your windshield or when a ribbon of fog sits above a stream in the early morning, "What's going on here?" When earthworms appear on the sidewalk after an evening rain, "What's going on here?" When you come upon a cluster of oak seedlings in a pine forest (with no big oak trees anywhere in sight), "What's going on here?"

Though you probably don't consider yourself a scientist, you are engaged in the four-step *scientific method* every time you make an *observation* that elicits a *question* that then prompts you to come up with an explanation (*hypothesis*) that you can test via *experimentation*. For example, suppose you want to enter a building. You see (observation) that there is a door and you wonder (*question*) if you should pull it or push it (alternative hypotheses). You decide to push (experiment 1), but the door remains stationary. Then, you pull (test 2) and the door opens. In quick succession, you enacted the four-step scientific method going from observation to question to hypothesis to experiment. In this sense, each of us is a daily practitioner of science.

Demystifying Science

Much of what we know about nature has been discovered by naturalists and field biologists. To illustrate and demystify how this discovery process works, suppose that one summer afternoon you see several orange-and-black monarch butterflies moving about in a field and you decide to spend some time observing these butterflies.[5]

Being a creative soul, you use your imagination to shrink yourself down to the size of a sand grain—small enough to fit on the head of one of the monarchs, right between her two antennae. This idea may sound a bit ridiculous, but scientists often exercise their imaginations in just this way—it's called practicing *umwelt*—knowing that it often takes a change in perspective to begin to understand the lives of other organisms.

Back to our monarch: It is a summer day and from your little perch between her antennae, you observe that as she flies about, she is laying eggs, each no bigger than a pinpoint. Although dozens of plant species exist in the meadow, you note that this monarch lays her eggs exclusively on milkweed plants. This observation prompts you to ask, How does she distinguish milkweeds from the scores of other plants in the field? Does she use sight? Smell? It turns out, based on field research, that monarchs use scent receptors on their antennae to smell their way in the general direction of milkweed plants. Then, once they land on a candidate, they use taste receptors located on their feet to verify if the plant is, indeed, a milkweed. As you continue your journey, you note that this monarch lays her eggs one at a time, often depositing just one egg per milkweed plant. Observing this, you wonder why the monarch spreads her eggs around so much. After all, wouldn't it be easier for her to simply lay all her eggs in one place and be done with it? But, with further observation, you discover that each caterpillar that hatches from a monarch egg gobbles up many leaves, and the supply of leaves on individual milkweed plants is limited. Hence, you hypothesize that by spreading her eggs among many milkweed plants this monarch mother ensures that her babies—the caterpillars that hatch from her eggs—will have enough food to grow to adulthood.

DEVELOP YOUR *UMWELT!*

The more time we spend in wild places—listening, observing, and paying attention—the more opportunities we have to experience the world from the perspective of other organisms. One of my early experiences with this occurred back in my graduate-school days. An acre of forest had been leveled and scraped away by a giant bulldozer, and a week later I was on my hands and knees searching to see if any plants had been able to colonize this scarred land surface. To my delight I did find some tiny seedlings here and there, and I marked each one with a toothpick. But the following week, when I returned to check on those seedlings, they were gone. I wondered what had happened. Hearing my question, a mentor challenged me to put myself in the place of a newly germinating seedling—in other words, to use my imagination to shrink myself down to the size of a tiny seedling trying to gain a foothold on sun-scorched bare ground. As I descended into that microworld, I felt the hot sun sucking the moisture out of my delicate leaves. I longed for relief and visualized a heavy rain falling. But when a rain came, it didn't bring relief; instead, it brought more trauma. With my new *seedling eyes*, the individual raindrops came crashing down, slamming into the soil, splashing my tender leaves with clay particles and creating minicraters that exposed my fragile roots to the burning sun. In this visualization, I was attempting to imagine the world as a seedling might experience it. There was no word for this concept until German biologist Jakob von Uexküll invented the word *umwelt* to signify the world around a living being as that creature experiences it.[6]

If you spend the time to develop your own *umwelt* capacities, you will gradually come to understand that each organism perceives the world with its own specific set of sensory tools. And then, the next time you take a stroll through a patch of forest, you will understand that you are experiencing only a sliver of the fullness of the forest. The doe, hidden from sight, will be experiencing that forest in different—but no less palpable—ways than you. So it is for the wood thrush, the earthworm, the moth, and the soil bacterium—each experiencing a distinct face of the forest. In truth, the forest is not one world; it is a rainbow of worlds. Endeavoring to experience the *umwelt* of other creatures is a wonderful initiation to the web of life, as well as an invitation to expand both our awareness and compassion.

You also observe that the monarch usually lays her eggs on the underside of leaves (secured in place with a gummy secretion), rather than on the more easily accessible upper surface. Wondering why and using nothing more than common sense, you first hypothesize that laying eggs on the bottom side might keep eggs from drying out, thereby increasing hatching success. Alternatively, you consider the possibility that bottom placement might render eggs less conspicuous to other creatures that could eat them. You decide to test these twin hypotheses by going on a search for milkweed eggs. Eventually you manage to locate one hundred milkweed leaves each with one egg on the undersurface. Then, you tag half of these leaves and leave them as is. As for the other fifty leaves, you rotate each one 180 degrees so that the egg on the undersurface now faces up. Then, by comparing the fate of the eggs over several weeks (e.g., dried up? eaten? hatched?) on the rotated versus nonrotated leaves, you gather the necessary data to test your hypotheses.

More observations in the milkweed field lead to more questions. For example, curious about the chemical content of milkweed leaves, you tear the edge of a leaf and see a milky latex fluid emerging from the wound (thus the name milkweed). Later, on the Internet, you read that biochemists have found that this milky sap contains cardiac glycosides—toxins that cause heart paralysis and a severe vomiting response in vertebrates. You and I would sicken if we ate a milkweed leaf, and, yet, monarch caterpillars eat milkweed with impunity. The reason, you learn, is that monarch caterpillars have evolved the biochemical capacity to excrete the milkweed's toxins—something that very few other species are able to do. But wait! Why do monarchs go to all the trouble of ingesting a plant loaded with toxins when there are other much less toxic plants to choose from? In pursuing an answer to this question you learn that monarchs are not only able to excrete toxic milkweed glycosides, they can also store these toxins in their tissues. In fact, biologists have discovered that the concentration of these toxic glycosides is higher in monarch tissue than in the milkweed leaves that they eat. Because of the toxins sequestered in their tissue, birds hunting for food learn to avoid monarch caterpillars. In fact, monarch caterpillars are brightly colored, presumably so that birds that make the mistake of sampling them once will remember their coloration and avoid them in the future. Indeed, researchers have found that blue jays, offered a variety of caterpillar types, quickly learn to reject the monarch caterpillars.

OK, let's take a step back. So far we have seen how observations, questions, hypotheses, and simple experiments have shed light on the biology of the monarch butterfly. But there is more to this story. Imagine, now, that the female monarch that you have been accompanying has finished laying her eggs. Her life is almost over. You sense this, and so, as she lays her last egg, you slide down from your spot between her antennae and nestle in beside that last egg. Eventually a tiny caterpillar hatches and commences to nibble away its egg case. Then, it proceeds to consume the fine hairs on the bottom of the milkweed leaf and then the leaf tissue itself. After a couple of days of feeding, the caterpillar becomes confined by its cuticle and undergoes its first molt.

Continuing to feed, this caterpillar goes through five molts over a three-week period, and then, big and full, she affixes herself to a branch and forms a mummy-like crucible within which her body is deconstructed and then reconstructed into an adult monarch butterfly.

When this adult emerges from her chamber, you are curious to learn what's next, and so, once again, you shrink yourself down and find a perch between her antennae. Imagine holding on tight as she spirals high into the sky in a courtship flight. You are still with her when she settles down in the late afternoon with a male suitor. And you watch as the male places a packet of sperm, complete with nutrient supplements, into a special receptacle inside her reproductive tract. The next day you accompany her as she begins to lay eggs, releasing sperm to fertilize each egg just before it is laid. This entire monarch life cycle is repeated several times during the summer, and then the weather turns cold. This weather change prompts a new question: How do monarchs survive the winter? Many insect species survive winter as eggs or as encased pupae, neither of which is the case with monarchs. So what do monarchs do? For many years it was a complete mystery. Some researchers, like Robert Michael Pyle, tried to literally follow monarchs on the ground as winter approached. For fun, put yourself in Pyle's position, a young man filled with curiosity, ambition, and boundless energy:

> I began at the northwesternmost breeding grounds, in the Okanagan Valley of British Columbia. My plan was to locate monarchs at their nectar sources and overnight roosts, then follow them as far as I was able on foot. I'd catch and tag them when I could. And when they rose and flew away, as they always would, I'd take their vanishing bearings—the direction in which they disappeared—as my running orders. I'd follow in the same direction, as much as the topography would allow, until I found more habitat, more monarchs . . . and then do it all again, as far as I could.[7]

As an alternative to Pyle's ground-tracking approach, Dr. and Mrs. F. A. Urquhart of the University of Toronto decided to mark monarchs. They did so by affixing tiny tags (like the self-adhesive price tags used in grocery stores) to the wings of adult monarchs born in the fall. Each tag had a number and a mailing address signaling where to send the monarch if it was found. So, for example, if one of the monarchs tagged by the Urquharts in Michigan was hit by a car in Arkansas and if the driver later discovered this tagged monarch on the car's grill and sent it back, the Urquharts would know that at least one of the Michigan monarchs went as far south as Arkansas. Using both ground tracking and tagging, researchers have slowly pieced together the migratory route and destination of the monarchs. Their winter roosts, it turns out, are in a mountainous region sixty miles west of Mexico City. There, monarchs, by the millions, settle down for the winter (figure 3.1).

Finding the monarch's winter roost in faraway Mexico raised yet another question: How does this delicate creature manage to fly several thousand miles from the United States to Mexico each fall? If they had to fly nonstop, they would starve to death. The solution, entomologists have discovered, is to feed along the way, following a trail of nectar-producing flowers. In fact,

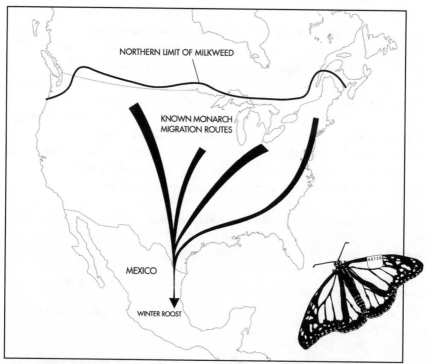

Figure 3.1. Monarch butterfly migration routes. Note monarch butterfly wing tag. These tags help scientists determine the monarch's migratory movements.

you can, if you choose, observe monarchs feeding and you don't have to shrink yourself to a sand grain to do it. You simply have to find a monarch in the early fall and follow it (just as Robert Pyle did) as it flies from flower to flower in a nearby field. When the monarch lands on a flower cluster you will actually be able to see it suck up nectar with its long straw-like tongue. And, if you have a sharp eye, you will see that pollen often becomes stuck to the monarch's legs as it maneuvers about on a flower head. As the monarch flits from flower to flower it carries this pollen to other flowers of the same species. So it is that in the process of feeding, monarchs also "give back" to the plant communities that they visit by serving as pollinators, ensuring seed and fruit production.

Remarkably, researchers have discovered that monarchs actually gain weight during their trip south by converting flower nectar to fat that they then store in their abdomen. These accumulated fat reserves are necessary for monarch survival during their three-month stay in the Mexican mountains where little food is available for them. At their winter roosts, the monarchs cluster on the branches of fir trees. The dense fir canopy protects them from freezing while keeping them cool enough to remain semidormant. Come the end of March, the monarchs rouse themselves and return north.

This final part of the story presented researchers with still another puzzling question: Do the same adults that spend the winter in Mexico fly all the way back to their home grounds in the north? The answer turns out to be "no." Biologists have learned that the adults leaving Mexico settle down to mate in the southern United States as soon as they find freshly sprouted milkweeds. After laying their eggs, these adult monarchs finally die; and it is their offspring that continue the journey north.

It wasn't trailing or tagging that resolved this mystery; rather, it was chemical fingerprinting. By analyzing milkweed chemicals present in the bodies of adult monarchs arriving to the northern United States and Canada, researchers were able to tell where these adults fed during their milkweed feeding larval stage. This was possible because there are scores of distinct milkweed species in the United States, each with its own limited geographic range and unique biochemical fingerprint. So it is that monarchs flying about in Canada in June have telltale chemicals in their body tissue unique to the southern U.S. milkweed species that they fed on when they were still caterpillars.

More recently the Internet has begun to play a role in providing access to monarch movements. Through "Monarch Watch" (www.MonarchWatch .org) and Journey North (www.learner.org/jnorth) anybody with an interest in monarchs can pass on information about monarchs' migratory movements. For example, monarch watchers in Mexico might send out the word in the early spring that the monarchs are beginning to leave their winter roosts. Then, observers to the north might begin to post messages—for example: "Monarchs just arrived here," "monarchs laying eggs," "monarch caterpillars more abundant than usual," and so forth.

The more researchers investigate the ecology of any particular species, the more they are presented with new questions. In the case of the monarch, a question that currently baffles biologists is: How do the monarchs that hatch in the fall in North America find their way, year after year, to the same winter roosting sites in Mexico, given that these monarchs have never been there before?

In sum, the monarch butterfly is just one species—one strand in the web of life. In reading this passage, you were, in effect, pulling on this strand, joining me in making observations and asking questions and, in the process, discovering that the monarch *strand* is connected to other strands—other species—each a part of the community of life.

FOUNDATION 3.2: RELATIONSHIPS ARE THE BASIS OF LIFE'S WEB

One of my students, Mike, knocked on my office door not long ago with a question. There was a walnut tree in his backyard that was bearing a heavy crop of nuts. Climbing this tree, Mike was surprised to find walnuts sitting in the crooks of several of the tree's branches. Mike thought it unlikely that walnuts were simply falling into the branch crooks and wanted to know what I thought. I didn't know the answer and suggested that he spend some time observing the tree at different times of the day, paying special attention to any animal visitors. The next day Mike was back in my office with a big smile. At dusk he had seen a squirrel lodging walnuts in a crook in a walnut branch, presumably for later consumption.

Mike, in effect, had found a web strand—that is, a connection between a squirrel and a tree. Such connections are everywhere, but to see them takes patience. In Mike's case, it took taking the time to sit quietly and observe.

Relationships in a Patch of Goldenrod

As with Mike, the opportunity to learn about nature is close at hand provided we are willing to pay attention. After all, our bodies are continually bombarded with electromagnetic waves (light), vibrations (sound), gases (odors), and infrared radiation (heat); and our sense organs translate this information into colors, sounds, objects, smells, and sensations. Oftentimes, though, our tendency to immediately categorize what we see interferes with our ability to truly experience the life around us. For example, you may see a sunflower and think, "There is a sunflower," but, in fact, you may only be seeing your memory of "sunflower" and not the real thing. But suppose that you resolve to really take in that particular sunflower. You see that the flower head—a complex medley of yellows and browns—is composed of a multitude of individual flowers, some fresh and open, others old and partly shriveled. Bringing your other senses into play, you gently move your fingers back and forth over the short hairs that adorn the sunflower stem; placing a petal on your tongue, you experience its taste; listening, you hear rasping as the wind rustles the sunflower's leaves; breathing in the subtle fragrances of the flower head, you are aware that molecules from this sunflower are merging with your olfactory cells, allowing you to literally take the odor of this sunflower into your body. Indeed, using one's senses to take in the world is a gateway to relationship.

With this in mind, each September I take my students to a meadow filled with wildflowers. When we arrive, I give each student a pocket-sized magnifying glass. Then, I instruct them to explore a patch of goldenrod, guided by my favorite question, "What's going on here?"

My students are initially drawn to the butterflies and bumblebees foraging on the goldenrod blossoms. They watch as these insects slurp up nectar. Some even note that in the process of foraging, these insects (like the monarch butterflies described above) pick up sticky grains of pollen, and carry this pollen from plant to plant, helping to ensure pollination (figure 3.2).

Looking more closely, students discover that pollinators are not the only insects at the goldenrod's floral banquet. There are also predators such as voracious ambush bugs that dismember their prey and tiny yellow crab spiders that lie in wait, camouflaged among the blossoms, ready to immobilize an unwary pollinator and suck it dry.

Below the flowers, along the stem, some alert students discover another drama being enacted: Aphids are sticking their needle-like proboscises into the goldenrod stems and extracting sugar-rich sap. But there is a problem. Plant chemists have determined that the goldenrod sap is low in the precious element nitrogen that aphids covet for protein synthesis. Hence, to satisfy their nitrogen needs, the aphids must suck up large quantities of the goldenrod's sugary sap. After they extract the nitrogen, the aphids let the surplus sugars drip from their anuses, but this excess doesn't go to waste. Students watch with fascination as they observe ants drinking the drops

WHAT'S GOING ON HERE?

One of my most bizarre and baffling experiences in the wild occurred while canoeing with a lifelong friend down Ontario's remote Spanish River. Each morning we woke early and canoed our way south. In the afternoon we bathed in the clean, cool river water. One day after our swim, as we lay reading at river's edge, we noticed a large animal, several hundred yards upstream, clomping along in the river shallows. At first, it looked like a horse but we soon realized it was a moose—a young female coming straight toward us. We watched as she approached: three hundred yards, two hundred yards . . . the sound of her hooves on the river cobble. At one hundred yards, she disappeared into a wet meadow and then suddenly reappeared, bursting out of the undergrowth, just thirty yards upstream from where we sat. We saw welts from fly bites on her flanks and heard her labored breathing. Without pausing, she crossed the river on a gravel shoal and disappeared into an alder thicket.

What going on here? we wondered. Why was this moose in such a hurry? Why was she seemingly oblivious to our presence? We sat pondering what we had just witnessed. Then, in the same spot—far upstream—near where we had first spotted the moose, we sighted a lone white wolf moving with equal purposefulness. Like the moose, this wolf also disappeared into the grassy marsh slightly upstream of us. Anxious to follow its movements, we scampered up onto a ledge. But we had no sooner gained stable footing than this very wolf appeared on that very same ledge, just ten paces in front of us. The instant she saw us, she spun around and vanished into the underbrush.

What was going on? Later, I learned that wolves sometimes track moose for days in hopes of eventually inflicting a fatal wound. Did this wolf pick up the moose's track again? I don't know. I do know that for a brief moment my friend and I were in an unfamiliar and very exciting part of life's web.

of sugary fluid leaving the aphids' anuses. Just as humans milk cows, ants, in effect milk aphids. This a boon for the ants, but the aphids also benefit insofar as the ants ward off ladybugs and other potential aphid predators. This service is important because a single ladybug, in the absence of any constraints, can easily consume fifty or more aphids in a day.

Further down on some goldenrod stems, students discover strange, marble-sized globes of tissue—goldenrod galls. I cut open one of these galls and show students the writhing larva inside. I then explain that if they had visited this field in the spring, they could have seen an adult fly known as the "goldenrod gallfly," developed from just such a larva, emerging from a ball of goldenrod tissue, just like the one in my hand. The adult gallfly would

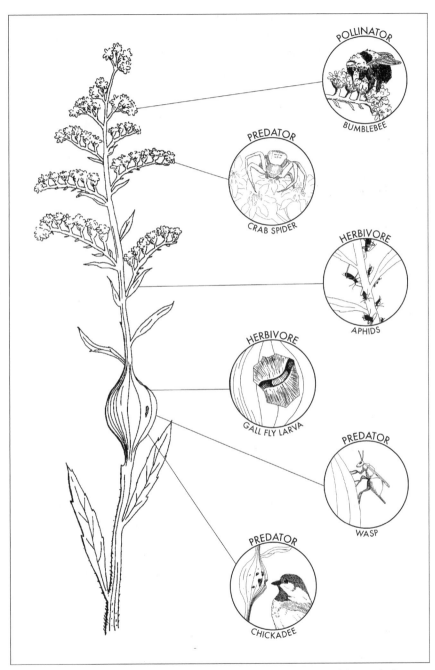

Figure 3.2. Pollinators, herbivores, and predators on a goldenrod plant.

have flown around laying eggs, one by one, on the tips of newly emerging goldenrod plants. The larva from one of these eggs would have tunneled down into a young goldenrod stem and chemically induced the goldenrod plant to produce a marble-sized gall—again, just like the ones my students had discovered. Gallfly larvae live inside these protein- and carbohydrate-rich galls through the summer and early fall; then they enter a resting phase before emerging as adult flies in the spring to repeat the cycle. But researchers have discovered that the lives of these gallfly larvae are some-times cut short in summer by a certain wasp species that deposits its eggs inside the galls. These eggs hatch into voracious wasp larvae that devour the gallfly larvae. If gallfly larvae are fortunate enough to escape attack by wasp larvae, they might still get gobbled up in winter by hungry chickadees and woodpeckers that peck open galls and devour their contents.

Even after I tell students we are done for the day, some linger in twos or threes by patches of goldenrod looking at this mini-ecosystem with new eyes. Indeed, a wealth of biology is contained in a flowering patch of goldenrod; it is a minicommunity, replete with pollinators, predators, herbi-vores, symbionts, and parasites.[8] And it is much the same in any other patch of wild vegetation, once we develop the eyes to see it.

Mutualistic and Symbiotic Relationships

For much of the twentieth century, biologists tended to describe nature as a "red tooth-and-claw" struggle—a place of high-stakes conflict where only the strong survived. On occasion, instances of cooperation among species were noted, such as when a plover flies into the gaping mouth of a crocodile to rid the croc's teeth of leeches (a nuisance for the croc but a banquet for the plover!). But such stories were often presented as curiosi-ties or exceptions to the more general view of evolution as grounded, almost exclusively, in hardscrabble competition. In recent decades, however, this view has begun to change as scientists are discovering more and more in-stances of cooperation and mutual interdependence among species. Here are four examples showcasing the ways in which organisms—sometimes very different from one another—interact in complementary ways.

Blue Jays and Oaks

I never saw a connection between oaks and blue jays until I observed a jay eating a white oak acorn one morning in late September. Grasping the acorn with its feet, the jay hammered it open with its bill and consumed the nut. Meanwhile, higher up in the same tree, I spotted another jay swallowing entire acorns, one after the other. I learned later, reading reports in scientific journals, that jays pack whole acorns into their expandable throat cavities. After loading up, they fly hundreds or even thousands of yards to their breeding territories. When they land, the jays disgorge their acorns and bury them, one by one, by pushing back soil with their beaks and then hammering the acorns into the soil; they then finish the job by covering each acorn with leaves and other debris.

A single jay may bury four to five thousand acorns over a one-month period in the fall. When winter descends and jays no longer have access to insects and worms, these acorns become a kind of survival food. Remarkably, the jays memorize the location of buried acorns by their spatial relation to nearby objects (beacons), such as rocks or logs. In fact, scientists have shown that, under experimental conditions, birds are no longer able to correctly locate their acorn caches when experimenters intervene (unbeknownst to the birds), changing the positions of the bird's natural beacons.

This relationship between oaks and jays is mutually beneficial: It provides jays with food to survive the winter; and it also enhances oak reproduction because, over the course of the winter, jays only dig up a fraction of their buried acorn stock. The remaining acorns are frequently "planted" in favorable conditions for germination and growth.[9]

Ruminants and Bacteria

Not all collaborations are as readily apparent as that between oaks and jays. Sometimes one of the partners is invisible to the naked eye. This is the case with cows and bacteria. Scientists, studying the cow–bacteria partnership, sometimes create a surgical opening, called a "fistula," in the cow's side to study the cow's digestive processes. Imagine, if you will, reaching your gloved hand through a fistula and pulling out a handful of mush from a cow's rumen. On first glance, this mush would appear to be nothing more than partially digested grass. However, if you were to take a tiny pinch of it, dilute that pinch in a few drops of water, and then place it under a microscope, you would see an extraordinary community of bacteria—some ciliated, others flagellated. It is these bacteria that do the main job of grass digestion in the cow's rumen. Without these bacterial symbionts, cows would be unable to digest the sugar-rich cellulose of grasses. As microbial biologist Lynn Margulis explains: "The cellulose-degrading microbes, in a very real sense, are the cow. . . . No cow would be grass-eating or cud-chewing without their microbial middlemen."[10]

Flatworms and Algae

Once, years ago, I attended a lecture where the speaker asked the audience why there isn't a mammal that can both walk around like us **and** produce its own food through photosynthesis. Wouldn't it be convenient, he mused, if you and I had banana plants sprouting from our shoulders? Then, when we became hungry, you could just reach up and pluck off some food.

Although such talk appears to be the stuff of fantasy, some animals—such as the flatworm, *Convoluta roscoffensis*—actually do carry their food system with them. If you were to encounter one of these worms (which look like gooey green ribbons) in their natural habitat along the beaches of northwestern France, you might mistake them for some sort of seaweed. The worms' bodies are green because their tissues are imbued with cells of photosynthesizing algae. These algae are the worm's "bananas." In fact, the worms' mouths cease to function once the algae come on board. Instead, they rely entirely on the algae's photosynthetic products for sustenance. "Algae and worm make a miniature ecosystem swimming in the sun."[11]

Fungi and Tree Roots

Walking through the woods after a wet spell, it is not uncommon to encounter mushrooms. Think of them as fungal "flowers" because, like flowers, they are the fungus's reproductive structures. If you take a mushroom cap, set it on a piece of paper, and then lift up the cap a few hours later, a print of the mushroom's spores ("seeds") will be left behind. These spores are so light that they can be carried great distances by wisps of wind. Hence, the mushroom, with its mobile spores, is the fungus's way of reproducing itself.

Although an individual mushroom may be quite conspicuous, it represents only a tiny part of the fungus's body. In fact, mushrooms wouldn't exist at all if it weren't for their underground bodies, which consist of complex networks of tubular threads known as "hyphae." Fungal hyphae absorb water from the soil and release enzymes that decompose the organic debris that then serves as fungal food.

It turns out that many fungal species form symbiotic partnerships with tree roots. For example, oaks, pines, maples, and hickories have specific species of fungi that associate with their roots. The trees actually feed carbohydrates (created through photosynthesis) to these fungi. In return, the fungi, by spreading their hyphal net out from the tree's roots, dramatically increase the ability of tree roots to absorb nutrients from the soil.

These various symbiotic relationships are like marriages, with an important distinction—the partners are distinct species. We may never know the exact steps that led to the formation of the multitude of symbiotic relationships that now exist in the world; we do know that these relationships are among the most important in life's web. As renowned biologist Lewis Thomas has pointed out, "The urge to form partnerships, to link up in collaborative arrangements is perhaps the oldest, strongest, and most fundamental force in nature. There is no solitary, free-living creature, every form of life is dependent on other forms."[12]

Humans Are Entangled in Life's Web

We humans need look no further than our own bodies to discover that we are deeply entangled in life's web. Indeed, your body, just like mine, is a swirling, pulsating ecosystem filled with other organisms. For example, each of us has several distinct species of bacteria living in our mouths. Some of these bacteria species are adapted to the environment at the tip of the human tongue; some prefer the back of the tongue; others, the cheek pouches; yet others, the plaque on our teeth. Another raft of bacteria make their home in the human gut. Some of these microbes help in the digestion of our food and in the manufacture of essential nutrients like vitamin K; others help tune our immune system, keeping harmful microbes in check.

None of our houseguests is better known than the bean-shaped bacterium *Escherichia coli*. Indeed, there are so many *E. coli* in your gut that they would stretch from Boston to San Francisco, if laid out end-to-end. Although minuscule, *E. coli* is a fully functional creature: It moves using whiplike appendages; it feeds; it reproduces; and it recognizes and communicates with its own kind.[13]

Bacteria also inhabit the outside surfaces of your body; there are hundreds of thousands per square inch on your legs, arms, trunk, and face. What are they doing? They are eating; they survive by feeding on the mix of salt, water, nitrogenous compounds, and oils that continually seep from the human body. And it's not just bacteria that live in and on us. Indeed, the chances are very good that there are scale-covered, wormlike mites known as *Demodex folliculorum* feeding in the follicles at the base of your eyelashes right now. These mites use their needle-like mouthparts to scarf up secretions and dead skin associated with human eyelash follicles. At maturity, the mites mate and then, soon after, the females deposit their eggs in eyelash follicles. The whole cycle, from egg to adult, takes about two weeks.

MADURA FOOT

The idea that our bodies serve as habitat for other creatures is something that I accepted intellectually, but I didn't fully experience it in a bodily way until I discovered *pulgas* living inside me. It all began with a slight itch on the heel of my left foot. I was living in a village along the Rio Negro in Amazonia. As the itch intensified, a white spot, about a quarter-inch across with a dark center, developed on my heel. Soon, I realized that it wasn't just one itch but several and that both of my feet were involved. I decided to consult *Merck's Manual*, a kind of bible when it comes to diagnosing tropical ailments and diseases. As I sat hunched over a table reading by candlelight, I ruled out one thing after another, but then I came to *madura* foot. "Madura" means ripe or rotten in Spanish; and *madura* foot is a fungal disease that transforms the human foot into a mass of oozing sinuses and rotting flesh. It seemed plausible, based on the book's description, that I was in the early stages of this disease. As I was fretting over my fate, my field assistant, Getulio, stopped by. He took one look at my feet and said, *"Pulgas."* Then, he put a match to a needle and dug open one of the white spots. Soon I could see a cluster of very tiny eggs. Getulio explained that these eggs were deposited in my foot by a common tropical flea that burrows into human feet. The eggs hatch; the larvae feed; then they exit the foot as adult fleas. Although I found this somewhat disconcerting, I was relieved to be habitat for a flea rather than for a fungus that could turn my feet to mush.

Fortunately, this creature's digestive system is so efficient that it doesn't need an anus. So you don't need to worry about getting poop in your eyes![14]

The big message here is that each of our bodies is an ecosystem filled with diverse habitats that offer shelter and food to a wide array of creatures. These organisms are everywhere, filling up and defining our bodies. In fact, it is estimated that the number of bacterial cells taking up residence in the human body exceeds by tenfold the number of cells of the human body proper.[15] Equipped with this awareness, it becomes more difficult to see ourselves as separate entities. The truth is that thousands of web strings emanate from each of our bodies, both outward and inward, to the multitude of earthly creatures that depend on us for their sustenance and well-being.

A Thought Experiment

Just as many species depend on us, we depend on many species in a dizzying assortment of ways. Biologist Gretchen Daily suggests a thought experiment (inspired by John Holdren) to fully appreciate the scale of our dependence on Earth's biota.[16] Start by imagining that you have been put in charge of an expedition that will create an independent, self-sustaining civilization on the moon. To make the challenge a bit less daunting, assume, first, that the moon has a friendly atmosphere just like our own with lots of oxygen; second, that the moon's surface is composed of pulverized rock (the basis for soil); and further, that it has water. Your specific job is to decide which organisms you will need to bring from Earth to ensure that your expedition can thrive on the moon for tens of thousands of years into the future. The technical commander assures you that your spaceship is very large so that you will have ample room for whatever you need to bring.

Food is the first thing that comes to your mind. So you set about gathering seeds of grains (corn, wheat, oats, rice, barley, etc.), vegetables (carrots, onions, kale, potatoes, squash, etc.), fruits (bananas, oranges, apples, grapes, kiwis, mangoes, etc.), and spices (thyme, rosemary, black pepper, ginger, garlic, etc.). Then, of course, you will need coffee and tea, as well as hops and yeast (for beer). And, oh yes, you will also want cane for sugar and cacao for chocolate. And finally, you mustn't forget to bring along spores from an array of mushroom types.

In the process of making your list and gathering the necessary fixings, you come to realize, perhaps as never before, that humans rely on hundreds of different species of plants. And not just plants! You remember that you will need to take chickens, goats, cattle, sheep, and turkeys, as well as water tanks filled with trout, salmon, shrimp, flounder, clams, and oysters, and on and on.

At last, you think that your job is done, but then you remember that you will need to make clothes; and so you add cotton and hemp plants to your list and perhaps silk-producing caterpillars and various plant species that provide dyes. You will also need to build things from wood and to create paper, so onto your list go dozens of different tree species. "There, that should do it," you say to yourself. But then an alarming thought occurs to you: What are you going to feed all your goats, chickens, cattle, and sheep? So you add alfalfa, forage grasses, nutritious herbs, and hardy shrubs to your list.

At this point, you are pretty sure that you have everything covered, but to be certain, you decide to imagine yourself actually living on the moon. There you are with all your seeds; but as you go to plant them, you realize that the soil is sterile—barren of the bacteria and fungi that are essential for nutrient cycling. And so you collect tons of living soil to bring along on your expedition, in hopes that the organisms in this soil will reproduce and spread.

As you imagine your crops and fruit trees growing, you realize, with a jolt, that you forgot the insects, birds, and bats that will be necessary to pollinate your food plants. So you set about compiling a list of all the pollinator species you will need. Once finished, you begin to wonder what your various pollinators will survive on during those times when their preferred crops are not supplying pollen and nectar. This worry sends you in search of wild (noncrop) plant species that can supply the various pollinators with food throughout the entire lunar year.

Finally, one night, just a month before your scheduled departure, you awaken from a nightmare. In your dream, insect pests are destroying the crop plants. You realize that the domestic animals and plants that you will be taking to the moon will undoubtedly harbor some microbial and insect pests. Since synthetic pesticides are not effective in the long run, you will need to take along all of the natural enemies of all imaginable crop and livestock pests.

Now, as you are beginning to really think for the long term, you realize that, given enough time, environmental conditions (e.g., climate) will likely change on the moon, just as has been the case on Earth. Hence, it won't work to take just a single variety of each species of plant and animal. Instead, you will need to have as many strains as possible to ensure that the raw material (genetic variation) for evolution is present.

This thought experiment reveals that to create a suitable environment able to sustain one species (human beings), we need a whole community of species. What's more, it turns out that even with all our seemingly sophisticated knowledge, we don't even have the scientific knowhow to create a self-sustaining ecosystem here on Earth (much less on the moon!). Our shortcomings in this regard were illustrated by the recent failure of the so-called Biosphere 2 experiment.[17] Biosphere 2 was designed to function as a completely self-contained artificial ecosystem. It was constructed under a transparent, airtight bubble in the Arizona desert. Three acres in size, this $200 million–plus unit featured zones of agricultural land, rain forest, desert, and savanna, along with wetlands and a tiny ocean with coral reefs. Biologists were hired as consultants to design the various subsystems and to determine the array of plants, animals, and microbes to be incorporated into this minibiosphere.

After everything was set, eight people were sealed into the unit. This team intended to stay for two years, but the experiment had to be terminated early in part because the atmosphere in the unit deteriorated; the oxygen concentrations dropped from a normal 21 percent to 14 percent—an oxygen level typical of elevations of 17,500 feet—and nitrous oxide concentrations rose to levels that can impair brain function. Meanwhile, nineteen of twenty-five vertebrate species in the unit went extinct along with all the pollinators, thereby dooming most of the plant species to eventual extinction. In addition, algal mats polluted the water while ants, cockroaches, and katydids underwent population explosions.

The eight Biospherians learned a basic lesson the hard way: Humankind is not only dependent on thousands of species in direct ways for food and fiber, but also on tens of thousands, perhaps millions, of other species for basic life-support services—that is, the services that maintain healthy ecosystems, such as nutrient cycling and air and water quality maintenance. Whether we know it or not, we are deeply insinuated into the living body of Earth and our knowledge pales in comparison to the complexity of nature.

STEPPING BACK TO SEE THE BIG PICTURE

> In reality there is a single integral community of the Earth that includes all its component members whether human or other than human. In this community every being has its own role to fulfill, its own dignity, its inner spontaneity. . . . Every being enters into communion with other beings. This capacity for relatedness, for presence to other beings, for spontaneity in action, is a capacity possessed by every mode of being throughout the entire universe.
>
> —Thomas Berry[18]

Though we have been conditioned to see ourselves as fundamentally separate, news arrives almost daily revealing novel ways in which we are, actually, connected to each other and to all of life. In fact, recent research reveals that we are influencing each other all the time. For example, as you read this, someone in a nearby room (assuming you are inside) could be subtly affecting your body temperature and your level of calmness (versus anxiety) without you even being aware of this. This has actually been scientifically documented in a simple experimental setup involving one person acting as a "Receiver" and a second person acting as a "Sender." Sender and Receiver are isolated from each other in separate soundproof rooms so that they cannot communicate by any known sense—that is, the possibility of communicating via hearing, sight, taste, smell, and touch are eliminated. The Sender's job is simply to focus mental attention on the Receiver during randomly selected thirty-second intervals. When attention is not being focused on the Receiver, the Sender simply relaxes. Meanwhile, the Receiver is instructed to simply keep his/her mind in an open state, avoiding any mental fixations. Before beginning, electrodes are placed on the Receiver's fingertips to monitor his/her emotional state. These electrodes measure changes in skin resistance (akin to a lie detector test).

The first studies of this type were conducted in the late 1970s at the Mind Science Foundation in San Antonio, Texas, and the results were highly significant, meaning that Senders did, indeed, influence the emotional arousal of the Receivers at a rate much higher than would be expected by chance alone. Since that time this same study has been repeated nineteen times in laboratories throughout the United States and Europe, and the overall result, again, has been highly significant.[19]

Just as humans sometimes tend to underestimate or ignore their capacity to influence each other, so, too, we easily discount the capacities of other life forms to "receive" our intentions. This was revealed in research conducted by William Braud at the Institute of Transpersonal Psychology in Palo Alto, California. Specifically, Braud wondered if less sophisticated forms of life, such as simple red blood cells, which do not, so far as we know, possess consciousness, could respond telepathically to human intention. The experimental setup was both simple and elegant. Red blood cells were placed in test tubes containing distilled water. Normally, red blood cells cannot survive in this environment; their cell walls collapse and the

---

JUST BECAUSE IT CAN'T BE EXPLAINED DOESN'T MEAN IT CAN'T HAPPEN!

Scientists, though they should know better, sometimes discount strange findings, like those described above, simply because there is no explanation for them within our current worldview. But it only requires a quick look at recent history to see that many phenomena that today we accept as commonplace would have been deemed impossible a few hundred years ago. For example: "Television and cell phones would have seemed miraculous to an eighteenth-century physicist, knowing nothing about electromagnetic fields. Hearing the voices of people far away would have seemed like the work of witches or the delusions of lunatics. But now this is an everyday experience, thanks to radio and phones. Likewise, hydrogen bombs would have been unthinkable for nineteenth-century physicists. In the age of steam and gunpowder, such devices would have sounded like an apocalyptic fantasy. And lasers would have sounded like mythic swords of light. They only became conceivable for twentieth-century physicists through the scientific revolutions wrought by relativity theory and quantum theory."[20]

cell contents go into solution. The amount of red blood cell damage can be determined by monitoring light transmission using a spectrophotometer. Light transmission through test tubes with intact red bloods is much less than transmission through tubes where the red blood cells have split open spilling their contents into the water.

Braud recruited thirty-two people for this experiment. Each participant had the specific task of telepathically coming to the aid of red blood cells placed in distilled water in ten test tubes. For each participant ten additional test tubes with red blood cells served as controls; the cells in these tubes had to fend for themselves—that is, they were ignored by the participants. All tubes were placed in a room separate from where the participants were located. Astonishingly, Braud found that the participants were able to significantly reduce the amount of red-blood-cell damage in those tubes that they directed their attention toward, relative to the control tubes. Moreover, they were even more successful in preventing cell death when the experiment was run with blood that they had personally contributed.[21]

There's more: There is now evidence that we humans are even able to influence the functioning of machines. The research comes from the work of Robert Jahn and Brenda Dunne at Princeton's Engineering Anomalies Research (PEAR) laboratory. In one carefully controlled series of experiments, these two scientists used a simple machine that generates a random set of zeroes and ones—like heads and tails. In a series of a thousand numbers, the machine randomly generates approximately 500 ones and 500 zeroes and the probability that the totals might differ from 50:50 (e.g., 490 ones vs. 510 zeroes) can be calculated with precision. When left to run unattended, the machine prints out a completely random list of zeroes and ones, just as one would expect. However, when the researchers invited human beings to sit in front of the machine and to use their thought and intention to bias the machine's output toward more zeroes or more ones, they found that in many cases the machine's output was no longer random. What's more, these Princeton "experiments have been reproduced by sixty-eight different investigators in a total of 597 experimental studies. It has been estimated that when the results of all these studies are included, the odds against chance being the explanation for [the anomalous results] are 1 in 10^{35} (i.e., 1 followed by 35 zeros)."[22]

An interesting side note to this Princeton work is that there is, apparently, considerable variation among different individuals in their capacity to influence machines. Those individuals who were most successful described their experience in terms of entering into resonance by merging or bonding with the machine. The very highest scores were achieved by couples who shared a deep love for each other and who worked in tandem in the presence of the apparatus. These scores were up to eight times higher than individuals who worked alone.

In interpreting these findings, Dr. Larry Dossey writes: "Proposing that things [e.g., computers] can behave well or badly, or that they have any behavior at all, sounds preposterous to anyone schooled in twentieth-century science. Yet *things* appear to be avid accumulators of significances, particularly when they are chronically exposed to humans. They appear capable of *catching* our feelings, resonating with our emotions, responding to our meanings, and obtruding into our lives, often when we least expect it."[23]

These new findings from science challenge the notion that there is such a thing as a separate, autonomous individual, suggesting instead that we are all entangled—humans and nonhumans alike—in ways that we haven't even begun to understand. In fact, research in the realm of quantum physics reveals that at the micro scale, separate particles can be made to act as if they are one and the same, even though the particles in question might be separated by hundreds or even millions of miles. If one particle of the pair is altered, its partner will be altered. The change is instantaneous, faster than the speed of light, absent any known exchange of information. This strange, seemingly mystical, interconnection is called "quantum entanglement." When particles are entangled, they cannot be described independent of one another, once again showing that relationship is knitted into the very fabric of the universe.[24]

Recognizing that the universe is grounded in relationship, the philosopher-scientist Arthur Koestler introduced the concept of the "holon."[25] Each whole thing within nature—whether a cell, an organ, a body, a lake— is a holon. In this conception, all holons are made up of smaller wholes as well as being a part of larger wholes. Koestler's elegant formulation reminds us that no living system can be entirely independent because each is, by definition, embedded in larger holons upon which it depends for its continued existence, as well as being composed of smaller holons. For example, a human being is a holon made up of smaller holons (organs, cells, molecules, atoms, and subatomic particles) and dependent for continued survival on larger holons (e.g., ecosystem, biome, Earth, Sun, galaxy, universe) within which he/she is embedded.

This idea of *wholes within wholes* offers a powerful metaphor for understanding how the world is put together. Indeed, the old metaphor, "world as machine," issuing forth from the Industrial Revolution, no longer fits our modern scientific understanding of the world. It was flawed from the outset insofar as it mistakenly saw all things as separate, isolated, autonomous. The holon now offers a welcome relationship-based understanding of reality.

WRAP-UP: EXPANDING ECOLOGICAL CONSCIOUSNESS

> When you really think about it, everything is connected—it's all con-
> nected.
>
> —Sally (Everglades woman)

Life's essence is relationship; Earth pulsates with relationship. Yes, that monarch butterfly laying its eggs on milkweed plants as well as that fox preparing
to pounce on a field mouse—all creatures large and small—are part of the
one true *worldwide web* of relationship. You and I are in this web, and this
web is in us. All life manifesting as *wholes* within larger *wholes*, containing
smaller *wholes* within! This is what our best science is now revealing to us.

But even granting our myriad interdependencies, there is still a strong
human tendency to regard our own species as separate from, and superior
to, all other life forms on Earth. This belief has been coined "speciesism."
However, just because humans happen to have big brains and the ability to
communicate complex messages and make tools does not necessarily mean
that we are *superior* to the millions of other species that inhabit Earth
alongside of us. All life forms, after all, have their unique capacities. Plants
have the ability to snag solar photons, speeding 186,000 miles per second,
and then to use those photons to transmute water and carbon dioxide into
food. I've never met a human who could do that! Or how about the lowly
horseshoe crab, a species that has managed to survive the rigors of Earth's
changing environments for more than a hundred million years? If longevity were the primary criterion for superiority, humans—who have not even
managed to survive for a million years—would be clearly inferior!

Even in the realm of cognitive intelligence, there is no guarantee that
humans are top dog. So far as we know, intelligence is correlated with
brain structural characteristics such as the area and layering of the neo-
cortex and the folding of the brain's cortical surface. In both of these
realms the brains of dolphins may surpass those of humans. The implica-
tion: We may simply not yet be smart enough to understand and appreciate the nature of dolphin intelligence.

In truth, we are not superior, nor inferior, to the rest of the family of
life. The process of evolution has produced Earth's *tree of life*. This tree
has myriad branches, each its own evolutionary line. The buds along each
of these branches are species. Our species, *Homo sapiens*, represents one
bud on one branch—no better, no worse than the millions of other buds
strung along the tree's other branches. It is the totality of species—the vast
range of life strategies and competencies—that constitutes the wonder and

magnificence of life on Earth and not the particular attributes of any one species. As we begin to comprehend our profound interconnectedness to the whole of creation, we awaken to the realization that all life is one—all made of the same stuff—all part of the same unfolding of intelligences—all interacting and fitting together like pieces of an enormous tapestry.[26]

This profound revelation—that we are not separate *from*, but part *of*, life's web—coming from both science and the world's wisdom traditions—has the capacity to awaken us to our true place in the Earth community. But for this to happen, it is not sufficient to understand, intellectually, our connectedness to Earth; we have to live this relationship, experience it, embody it; and this requires that we be willing to recognize and respect the denizens of the more-than-human world that we dwell among.

I was reminded of this when, on a recent hike, I stopped by a stream to have lunch and write in my journal. As I sat eating, a tiny yellow spider dropped down in front of me on a silken thread. I enticed her onto my pencil and then had a close-up look with my hand lens. She was magnificent, delicate, almost translucent—with a red spot on her abdomen. As I watched, she played out a silk thread from her spinneret, dropped down a few inches and then reeled herself back up to the pencil tip. Using the hand lens again, I was able to see fine bristles on her legs and the complex articulation of her mandibles.

Eventually, I coaxed her off the pencil and watched her scurry onto my jacket. A few minutes later I felt something behind my ear and instinctively rubbed it away. In that instant I remembered the little yellow spider—the creature that, a few minutes earlier, had enthralled me. It was too late. The remains of that spider were now smeared across my fingertips. Normally, killing a spider wouldn't have mattered so much to me, but this time, because I had closely observed this particular spider and had marveled at her beauty and life force, I felt a bond, a relationship; and so I experienced a measure of remorse. I suppose that this is the difference between intellectually recognizing that everything is connected, and actually having the visceral experience of this truth.

APPLICATIONS AND PRACTICES: CULTIVATING INTIMACY WITH EARTH

Intimacy = Into-me-see

What comes to mind for you when you hear the word, "vulnerability?" Does it conjure fear? Repulsion? Weakness? This was the case for me for many years and so I avoided putting myself in unfamiliar and potentially embarrassing situations. Now I see it differently. Being vulnerable means removing my masks; it means taking risks; it means mustering courage; it means being real. I now understand that being vulnerable can be a precursor to intimacy.

Cultivating Intimacy with Earth by Observing and Asking Questions

One of my favorite practices for cultivating intimacy with Earth is called Fifty Questions. All that is required is a nice spot in a meadow or a patch of forest or along a stream or in a local park. Find your spot and simply sit quietly for an hour, carefully observing the life around you: the smells, the quality of the air, the sounds, the insects, the birds, the plants. As you pay attention to your surroundings, questions will inevitably arise: What bird is that flitting about in the bushes? Does it have a nest close by? What is that ant carrying? Where is it going? Why is there moss on this rock but not on that rock over there? Why do the leaves of oaks, pines, and maples have different shapes?

Stay in the here and now, observing and writing down questions as they come to you. If you give yourself over to this experience, questions will arise—sometimes in torrents. You will also learn things about the way that you sense the world. For example, examining the kinds of questions you ask will reveal the senses that are most active for you, the colors that most attract you, the kinds of organisms that most garner your attention, and so forth.

After an hour, you will have easily accumulated fifty questions, perhaps even a hundred. Some of your questions will already have known answers, but it is likely that some will have never been asked, much less answered.

Creating a list of questions is just the beginning. Don't stop there. Become a citizen scientist and pick a question from your list that is especially interesting to you and then join your curiosity, creativity, and common sense to discover the answer to that question.

Cultivating Intimacy with Earth by Discovering Yourself in Clay

In our culture it is common to "construct" a self—based on kudos and resume lines—expressed as a personal narrative of our worthiness and value. But what if we didn't need to construct a self? Didn't need to prove that we are worthy? What if each of us simply understood that we already *are* good and worthy—that goodness is fundamental to our nature? That we don't need to earn it? Just feel what it would be like to live within such a new story of self—and from this foundation, ask yourself, Who am I? And, then, rather than relying on your powers of reason and discrimination (so-called left-brain attributes) to answer this question, call up your visual, imaginative, sensual, and intuitive capacities (i.e., your right-brain attributes).

One way to activate your right brain is to simply ask Earth the question, Who am I? This might sound far-fetched, but think about it: Each of us is constituted of Earth; we came out of Earth, and some day we will each return to Earth. So why not ask Earth?

If you are game, go out and get yourself a fistful of clay. It is easy to do. Simply dig down a half-foot below the topsoil and you will likely find a seam of rusty-red clay. If not there, construction sites and creek banks are good sources for clay.

Use your paws to dig up a clod and take it home with you. Then, when you are ready, pick up your clay, sensing the weight and texture and smell of it. If you need to, add some water to make it more pliable. Next, close your eyes and begin to explore your clay with your moistened hands, noting the contours, depressions and bumps, the wet and dry places, the warm and cool zones. Then, with your eyes still closed, begin to work your clay.

Leave your thinking mind behind and allow your hands to take the lead, molding the clay in ways that simply feel good. Keep your eyes closed.

As you sink deeper into this creative process, say to yourself: "This clay is me and I am creating myself." There is no hurry. Allow your hands to decide when you are really done. And only then open your eyes and behold your creation. Run your fingers over and around your clay form, both looking at it and touching it, attentive to any feelings that arise. If further sculpting is called for, go at it.

When you are finished, consider: What insights, longings, visions, and fears are embodied in my sculpture? What is this clay—shaped by me— telling me, revealing to me?

What new story of self is embodied in my creation? As you reflect, take to heart these words of the famous psychologist Carl Jung: "Often the hands know how to solve a riddle that the mind has wrestled with in vain." Be patient. Place your creation on a stand to dry. Look at it from time to time— for example, before you sleep, when you wake up. Trust that the body of Earth—your larger body—has things to teach you, a story to tell you.[27]

Cultivating Intimacy with Earth by Learning the Geography of Your Home Place

When someone asks where you live, if you are like many people, you probably respond by giving the name of your city or town along with your street address. And what you probably neglect to describe, in any meaningful way, are the physical and ecological characteristics of your home territory—for example, the lay of the land, the geological features, the common trees and birds and insects, the types of soils, the common weather patterns, the special smells and the seasonal sounds particular to your home place, and more. In truth, many of us have no idea where we really live. How about you? Here is a simple test you can take to gauge your intimacy with your home territory:

> It's easy to do. You simply invite someone to visit you who lives at least twenty miles away and who has never visited you before. You can give verbal instructions on how to get to your abode over the telephone, but the one rule is this: In your directions you may refer to *anything but* human artifice [i.e., no fair referring to man-made things such as roads, signs, houses, businesses]. Your job, then, is to guide your friend to your house by referring only to elements in the natural world. So it is that you may refer to hills, oak trees, the constellations of the night sky, the lakes or ocean shores or caves, the positions of the planets or any ponds, trails, or prairies, the Sun and Moon, cliffs, plateaus, waterfalls, hillocks, estuaries, bluffs, woodlands, inlets, forests, creeks, swamps, bayous, groves, and so on. Whenever your friend gets stuck, she is free to phone you for more directions, but the rule for her is that she must describe her location, again, without referring to any human artifice.[28]

Very few people in the United States today could pass this test; most Americans, it seems, have become "displaced" people. Yet, I strongly suspect that those Native Americans living here prior to European settlement would have had little difficulty passing this test.

It is by settling in and paying attention and communing with the land and water and sky of our home places that we come to know them. As our home places gather meaning, they become special, unique, sacred even. When such places are desecrated, we might actually feel like we have been personally violated. In this vein, when I ask people to describe the wild nooks and havens of their childhood—e.g., the spreading oak, the small cave, the hidden pond—a surprising number confide that their childhood sanctuaries have been snuffed out, replaced by highways, malls, parking lots, airports, and subdivisions.

Teacher and author Scott Russell Sanders, whose childhood home in Ohio was inundated when a river was impounded to create a reservoir, believes that our home grounds can be saved if we recognize them, speak about them, and celebrate them. We need, says Sanders, "new maps, living maps, stories and poems, photos and paintings, essays and songs. We need to know where we are, so that we may dwell in our place with a full heart."[29]

How to begin? Gather with friends around maps of your home territory. On these maps draw in where your water comes from, sketch the streams that flow through your bioregion, the principal soil types of your home place, the hilltops that offer cherished vistas, the fragile marshes and wetlands, the best fishing holes, the places where bald eagles have been sighted, the spots where rare orchids occur, the locales known for the biggest blueberries, the places where woodcocks perform their spring mating rituals. Sit together and share stories of these places; rescue the lore of the land. Make pilgrimages to these special places.

Cultivating intimacy with our home places is a lifelong practice. It is never too late to begin. Take the case of Steve Talbott. At age fifty-one, Talbott wanted to do more than identify the birds in his backyard. He wanted to make contact with them. So he sat by a patch of shrubs in his backyard offering seeds to the birds, acting as a sort of human feeding station. In a few days he had chickadees, juncos, titmice, and nuthatches feeding from his hands. Talbott is quick to point out that he had no special skills and that his approach was utterly mechanical. But he was, nonetheless, deeply moved by his experience: "All this has been an epiphany for me. . . . I suggest that a bird in hand—and a pine cone, and a rock, and a crawdad, and a snowflake—are the counterbalance we need if our alienation from nature is not to become more than the world can bear."[30]

QUESTIONS FOR REFLECTION

- It has been said that we are all scientists—all curious problem solvers. How might this be true for you?
- How might it change your awareness and consciousness if you centered your life on the question, *What's going on here?*
- How might experiencing the "umwelt" of other organisms create greater intimacy—greater connection—between you and Earth's web of life?
- What opportunities do you see in the place where you live to become a citizen scientist?
- How does the realization that your body is a swirling, pulsating ecosystem, providing habitat for millions of creatures, affect your identity?
- What happens for you when you allow yourself to consider that, contrary to appearances, there may be no such thing as a separate, autonomous individual—that the underlying quantum reality of existence is that all life is intimately entangled?

Part II

ASSESSING THE HEALTH OF EARTH

Unless we change the direction in which we are headed, we might wind up where we are going.

—Chinese Proverb

If I were to tell you that someone you don't know and will never meet was terribly sick, even dying, chances are you wouldn't care very much. And that's probably a good thing. After all, every day each of us is surrounded by people, both near and far, that we don't know—people by the hundreds of thousands—who are struggling, sick, sad. If we were to become personally upset every time we learned of a sickness or accident or mishap in the world, we wouldn't be able to function.

But what if you actually knew this person who was "terribly sick," and not just knew her, what if you loved her? What if this person was your best friend? How would you react then? Now, go a step further and imagine that this ailing person was a parent or spouse—that is, a person so close to you that you depended on him/her for your sustenance, your well-being? What then?

Finally, what if you were to learn that this isn't just a hypothetical situation, but the reality in which you—all of us—now live? Huh! Who is this ailing beloved, you may wonder? Most of us can't see her because she is so close at hand—so fully integrated into us and our experience of being alive that we are blind to her presence. The beloved I am referring to, of course, is planet Earth.

The thesis of part II of this book is that Earth—the planet that birthed us, nurtures us, shelters us, breathes us—*is* terribly sick and, in fact, has been ailing for a while now. As far back as 1992, a statement issued by 102 Nobel laureates in science along with sixteen hundred other distinguished scientists from seventy countries, read:

> We the undersigned, senior members of the world's scientific community, hereby warn all humanity of what lies ahead. A great change in the stewardship of the earth and the life on it is required, if vast human misery is to be avoided and our global home on this planet is not to be irretrievably mutilated.[1]

Yes, Earth is now in a state of indisputable decline. Atmospheric chemists report steady rises in greenhouse gases; agronomists tell us that soils are eroding in most places more rapidly than they are forming; human physiologists report increases in foreign, sometimes disease-causing, chemicals in our bodies; ecologists register the greatest spike in species extinction since the age of the dinosaurs; sociologists observe the breakdown of families and the dissolution of communities; while philosophers and religious leaders note the erosion of moral principles and the increasing alienation of humans from each other and the natural world. In short, these researchers and thinkers know that our Nobel laureates were speaking the truth when they told us that "vast human misery awaits us" unless a "great change in our stewardship of earth and the life on it" is forthcoming.

Planetary upset is the unifying concept that knits together the three chapters of part II. I invite you to join me in examining Earth as a medical doctor might—first pinpointing the signs of ill health and, then, looking below the symptoms to the underlying causes of Earth's breakdown.

4

LISTENING

Gauging the Health of Earth

We see quite clearly that what happens to the nonhuman happens to the human. What happens to the outer world happens to the inner world. If the outer world is diminished in its grandeur then the emotional, imaginative, intellectual, and spiritual life of the human is diminished or extinguished.

—Thomas Berry[1]

I live close to Spring Creek, a cold-water trout stream located in the Ridge and Valley region of central Pennsylvania. One afternoon, shortly after taking a faculty position at Penn State, I decided to take a hike from my office to the banks of Spring Creek. I planned my walk so that I would be able to visit two springs feeding the creek along the way. My first stop was Thompson Spring. It wasn't as easy to get to this spring as I had anticipated. I had to cross a road clogged with traffic, climb over a fence, and then navigate my way around a wastewater treatment facility. There at the back of the treatment plant was Thompson Spring, set in a little grove of trees. I stood spellbound, watching as water bubbled up into a lovely pool about ten feet across. But just thirty feet away, the crystalline water cascading out of this pool was overwhelmed by the voluminous effluent from the wastewater treatment plant.

Walking east from Thompson Spring, fifteen minutes later, I came upon Thorton Spring. Thorton bubbles out at the base of a forested hillside. It would have been a fine place for a picnic if it hadn't been for the strange smell. Thorton Spring, I soon learned, is located just below a chemical plant. For many years this facility stored its chemical waste in underground storage drums. The drums eventually developed leaks and two highly toxic chemicals, Mirex and Kepone, seeped into Thorton Spring and from there into Spring Creek.

As we saw in the previous chapter, everything is connected. This means that if you mess with something in one place, the effects will show up, eventually, in other places. As we will see in this chapter: Mess with Earth's forests and you affect Earth's bird life and much more; mess with Earth's oceans and you affect Earth's sea life and you and me, as well.

It is clear that we humans have been doing a lot of messing of late, but should this be cause for concern? This is a question that my environmental science colleagues and I have been devoting attention to. Like a medical doctor with a stethoscope to a patient's chest, we have our ears to planet Earth, listening attentively for signs of well-being, as well as danger.

FOUNDATION 4.1: LISTENING TO EARTH'S SKY AND LAND CREATURES

Suppose that when you were growing up there was a striking bird species—say the Baltimore oriole—that nested in your neighborhood. And suppose, further, that now, as an adult, you no longer hear or see this favorite childhood bird when you return home. If this were the case, you might ask, "Where have all the orioles gone?" And, then, building on what you have learned in the previous chapter about the scientific method, you might hypothesize that something about your neighborhood environment has changed, making it no longer hospitable for Baltimore orioles. This would be reasonable. Indeed, the idea of using birds to gauge the health of the environment is not a new one. Coal miners used to take canaries with them into the mines. If the canary stopped singing or, worse still, croaked, the miners cleared out quickly, knowing that carbon monoxide was present in dangerous concentrations.

Listening to What the Birds Are *Telling* Us

Since the middle of the last century, U.S. scientists in conjunction with everyday citizens have been using birds to gauge the health of Earth. Not just any bird species will do. Ideal candidates are migratory bird species—

WHY MIGRATE?

You might be wondering why so many species of North American birds migrate south each winter. The quick answer is to escape the cold, and there is some truth in this. But if these birds go south for warmth, why do they bother to return north in the spring? Why not just stay where it's warm? The answer is food. By returning north in the spring, migrants gain access to the wealth of food that they need for rearing their young. Indeed, with the onset of spring, insects hatch out by the gazillions in the north. These insect hatches are timed to correspond with the unfurling of new leaves—the primary food source for most insect larvae.

Imagine for a moment that you are a bird and you arrive to a forest with literally thousands of insect larvae feeding on every tree! What bliss! Now, imagine that you are one of those trees with freshly hatched caterpillars beginning to munch on your unfurling leaves. What terror! How grateful you would "feel" to the migrant birds that come north and, almost as partners, pluck off many of the caterpillars that are eating your "body." Of course, trees don't experience "terror" or "gratitude," nor do birds (as far as we know) experience "bliss"; but, certainly, reciprocity is a fundamental necessity for successful coexistence among species.

that is, species that spend spring and summer in North America (where they breed) and winter in warmer regions like Central and South America. Examples of migratory bird species include Baltimore orioles; red-eyed vireos; wood thrushes; scarlet tanagers; red-winged blackbirds; ovenbirds; and many species of warblers, hawks, and eagles, as well as numerous types of ocean shorebirds.

Collectively, migratory bird species dwell in a broad spectrum of environments. For example, some migrant species prefer forests, others grasslands, others marshes, others shorelines. Scientists reason that if such wide-ranging bird species are healthy, then it is likely that the many habitats and environments these species depend on are also in good condition.

Researchers have employed several tactics to assess the health status of migrant bird populations. One approach is to census birds in the same parcel of land year after year. For example, the migratory birds breeding in Rock Creek Park in Washington, DC, have been surveyed each spring for more than a half century. The census is conducted in the springtime by people who walk through this large woodland park, day after day, identifying all the birds that are nesting. Although this may sound tedious, it can be exciting to be out early on a spring morning, listening to birds sing and observing their courting and nest-building behaviors. However, in the case of Rock Creek, the enjoyment has been diminished because the census takers have discovered that the number of breeding migratory bird species in the park has dropped by one-third in recent decades.[2]

Of course, it could be that the migratory species that have disappeared from this park are now thriving in other areas of the United States. Indeed, it would be necessary to have a thousand Rock Creek–type studies, spread across the United States, before firm conclusions could be drawn. Fortunately, owing to a unique collaboration between scientists and citizens initiated by the U.S. Fish and Wildlife Service in the late 1960s, just such a large-scale annual bird survey does exist.[3] Indeed, each spring professional biologists plus highly competent citizen scientists conduct bird counts along more than four thousand prescribed transects (each transect several dozen miles long) spread throughout the United States and Canada. The results from this enormous undertaking corroborate the Rock Creek study—namely: More than one half of North America's forest-dwelling migratory bird species have experienced significant population declines in recent decades.[4]

Why the Decline?

Documenting declines in the populations of migrant bird species has been only one part of the research challenge; explaining the causes of this decline has been the other. One obvious place to look for causes has been the tropics—the wintering ground for most North American migratory birds. We have all grown up hearing about tropical deforestation, and surely a connection exists between the depletion of migratory bird populations and the depletion of their habitat in Central and South America.

But tropical deforestation and associated habitat loss is only part of the story. The birds also face threats in the north. For example, North American field biologists report that many migrant species are now having a difficult time raising young. The birds return to the north in the spring; they are successful in finding mates; the females lay eggs; but, in the end, the breeding pairs often fail to fledge young. Why? What's going on?

Some of the factors at play may be evident in our own backyards. For example, have you ever seen a cat stealthily stalking a bird? I have on several occasions, but I never thought much about it until I learned that there are an estimated 120 million *outdoor* cats in the United States (including both pet cats and feral cats). These cats, collectively, kill an estimated 1.8 billion birds per year, many of them migratory species. This comes to roughly 10 percent of all birds in North America.[5]

Feral cats have proliferated, especially in the open, suburban-type environments that have been rapidly spreading throughout North America in recent decades. Raccoons and opossums also prey on bird eggs and, like feral cats, their populations have also increased in concert with the spread of suburbia. Ecologist Dave Wilcove—noting that the continuous expanses of forest that once covered many regions of North America have been sliced and diced into a mosaic of small forest fragments—wondered if forest fragmentation is a factor in the decline in migratory birds. Specifically, Wilcove hypothesized that migratory birds, nesting in small patches of forest surrounded by open land, would be more likely to suffer egg predation by animals such as raccoons, opossums, and feral cats as compared to migrants nesting in nonfragmented, contiguous forest where feral cats, opossums, and raccoons are less common. Wilcove tested his hypothesis by placing wild-bird eggs in artificial nests and then scattering these nests in fragments of forest ranging in size from a few acres to a few thousand acres. He used the Smoky Mountain National Park, five hundred thousand acres of contiguous forest, to represent the natural condition before the onset of forest fragmentation in North America. When Wilcove went back to check all his nests, he found that only one nest in fifty had it eggs eaten by nest predators in the densely forested Great Smokies; but in the smaller forest patches, located in rural and suburban environments, egg predation was much higher, rising to 100 percent in some of the smallest forest fragments.[6] Thus Wilcove's study offers experimental evidence linking forest fragmentation to reproductive failure in migratory birds.

FOREST FRAGMENTATION AND LYME DISEASE?

At a time when we have been led to believe that humankind has won the war against infectious diseases, it is sobering to learn that since 1980 the World Health Organization has identified more than thirty new infectious diseases. Epidemiologists studying the outbreak and spread of new diseases sometimes link outbreaks to human-induced phenomena like forest fragmentation, industrial farming, and global climate change. Take Lyme disease. A half century ago it was virtually unheard of in the United States, but today it is no longer unusual. The insect that transmits Lyme disease to humans is a small tick that normally feeds on the blood of deer and mice—species that thrive in fragmented, suburban-type landscapes. In recent years their populations have been exploding; and, so have the ticks. It now appears plausible (although by no means certain) that the spread of suburbia (with its attendant landscape fragmentation and increases in mice and deer populations) has been a significant factor in the explosion of tick populations. For humans, sadly, more ticks translates to a greater probability of contracting Lyme disease.[7]

It turns out that there is more to this story of migratory bird decline than forest loss and nest predation. For example, it may surprise you to learn that radio and cell phone towers also take a heavy toll on migrating birds, especially insofar as these birds often fly at night and navigate toward light in foul weather. In an earlier age this would have simply meant flying toward the moon. But now, in the age of electricity, it often means flying toward the red glow of communication towers. When this happens, the birds' homing "magnet" easily becomes befuddled by the transmitters' electromagnetic field, often causing the birds to circle the towers and, in the process, to risk getting *sliced and diced* by tower guy wires. If you have the stomach for a grisly expedition, go out and inspect the base of a transmission tower after a foggy night in early spring and you will likely discover the carcasses of warblers, wood thrushes, red-eyed vireos, and many other migratory bird species. Though sad, it might not seem like a big deal until we stop to consider that there are now several hundred thousand transmission towers in the United States alone, collectively killing several hundred million migratory birds each year.[8]

This saga of migratory bird decline is also playing out in Europe and Africa. More than half of the hundred-plus migratory bird species that raise their young in Europe's spring, and then fly to tropical Africa in winter, are now experiencing population declines, presumably because of habitat alteration.

And as if things aren't tough enough, there is now evidence that climate change can be a further deterrent to successful breeding for migratory birds. For example, spring is now arriving several weeks earlier in many areas of the United States and Europe, and this is causing some insects to hatch earlier than usual—before the spring return of migratory birds. Thus, certain migratory birds now run the risk of arriving to their breeding grounds after the spring-insect food bonanza that they have been accustomed to rely on for raising their young.[9]

In sum, many migratory bird populations worldwide are in decline. Are cats to blame? How about raccoons? Is it the tropical deforesters? Or should we point our finger at the land speculators and developers who are slicing apart forests with roads, malls, and sprawling suburban housing developments? What about those radio and cell phone towers? And don't forget the burning of fossil fuels that contributes to global warming and with it the early arrival of spring in the north? In fact, these are *all* factors that are contributing, in some measure, to bird decline. There is no single *smoking gun*; human activities on a multitude of fronts are jeopardizing the well-being of Earth's migratory birds.

Bird Population Decline in a Larger Context

If you had walked into an elementary school classroom in the 1950s and called out the word "extinction," and then asked students for the first word that came to mind, chances are good that you would have heard just one word—dinosaurs! At that time extinction was regarded as something that happened far back in time. But call out "extinction" in a classroom today, and students, in addition to calling out "dinosaurs," are likely to volunteer the names of a raft of rare and endangered species: California condor, blue whale, snow leopard, Florida panther, monk seal, gorilla, gibbon, and on and on.

Biologists now concur that human activities are precipitating a mass extinction episode. Whether it is by clearing swaths of forest for farming in the Amazon basin or replacing a woodland with a subdivision in New Jersey or damming a river in Oregon, the outcome is the same: habitats of plant and animal populations are being destroyed. The mistaken belief that Earth is simply here to serve as a supply house for human needs leads us to see the forest as a "woodlot" and the stream as a "hydropower generator"; but, of course, forests and streams, first and foremost, serve as homes for other species. Just as we require space to live, so do they.

When scientists talk about the magnitude of the current extinction crisis, they often contextualize their remarks by referring to background rates of extinction. Based on the fossil record, species persist, on average, about a million years—some shorter, some longer—before going extinct. Hence, for each million species on Earth, only one, theoretically, goes extinct each year; this is the so-called background rate of extinction. But Earth is now losing species at a rate that is estimated to be somewhere between a hundred and a thousand times faster than this "normal" background level.[10] Data from the World Conservation Union's Red Book reveal that "39% of the mammals and fish, 30% of amphibians, 26% of reptiles and 20% of bird species" on Earth are now in danger of extinction.[11]

What does this mean for the future? Stuart Pimm, a renowned ecologist at the University of Tennessee, used a combination of field observations, published data, and computer modeling to address this question and concluded that "we might lose between a third and a half of the life on Earth as a consequence of our actions."[12] In the face of this prognosis we would be wise to do everything in our power to heed the words of the great American conservationist Aldo Leopold:

The last word in ignorance is the man who says of an animal or plant: "What good is it?" If the land mechanism as a whole is good, then every part is good,

SO WHAT IF WE ARE LOSING SPECIES?

Are the high rates of species extinction that scientists now report and project into the future really anything to be concerned about? After all, many species seem to do pretty much the same thing. Why do we need ten different tree species in a forest? Wouldn't one be good enough?

This is a good question, and here is one way to think about it. Imagine that, as you prepare to board an airplane, you notice a guy up on the wing busily popping loose rivets. You walk over and yell up, "What the heck are you doing?"

He replies nonchalantly, "Our airline company has discovered that we can sell these rivets for two dollars apiece and we need the money. If we don't pop rivets, we won't be able to keep expanding and offering our customers more services and flight options." If you were to encounter such a situation, you would be wise to head back to the terminal, report the rivet popper to the FAA, and book your flight with another carrier that was committed to passenger safety first.

The point of the story is that all of us, whether we choose to acknowledge it or not, are already riding on a very large carrier, *Spaceship Earth*, and we don't have the option to switch to another craft. The *components* of our "aircraft" are Earth's ecosystems, and our craft's "rivets" are akin to the myriad species living in Earth's ecosystems. It is probably true that removing a few rivets from the airplane wing—i.e., driving a few species to extinction—will not compromise the functioning of the airplane or the ecosystem. However, it might happen that the ninth rivet popped from a wing flap is critical and its removal will seriously jeopardize the aircraft's safety; just as removing a key species from a forest—such as a tree species critical to the nitrogen cycle—could undermine the health of the forest.[13] Indeed, the failed Biosphere 2 experiment described earlier reminds us that Earth's ecosystems are far more complex than we are presently able to understand.

whether we understand it or not. If the biota, in the course of aeons, has built something we like but do not understand, then who but a fool would discard seemingly useless parts? To keep every cog and wheel is the first precaution of intelligent tinkering.[14]

FOUNDATION 4.2: LISTENING TO THE EARTH'S OCEAN CREATURES

Early astronauts, looking back to Earth from space, observed what we land-dwelling humans often overlook—namely: Earth is mostly blue. These

Figure 4.1.   An ocean-centric perspective on planet Earth.

pioneers were almost brought to tears when they realized that Earth was the only thing in space with any color to it; everything else appeared black, white, or gray. Figure 4.1 offers an *ocean-centric* view of planet Earth, reminding us that Earth is predominately ocean, not land.

It is not just the vast extent of the ocean—occupying almost three-quarters of Earth's surface area—but, also, the vast depth of the ocean that is noteworthy. Yes, the ocean is not two-dimensional but three-dimensional, with an average depth of two and a half miles and extending down seven miles in some places. Therefore, compared to land, Earth's seas offer an enormous volume of living space—habitat—for complex life forms. Indeed, recent estimates suggest that more than 95 percent of Earth's habitable volume is found in the sea.[15] And, yet, if you are like most people your encounters with the ocean have been limited to the beach, where sea and land touch. Few of us have ventured out into the open sea, much less to the ocean depths.

Alien World or Cradle of Life?

The sea is unfamiliar territory for me. It is not my habitat. I had a visceral experience of this some years back when I was invited to go snorkeling for the first time along a coral reef in the Caribbean. Plunging into the water I was surprised to see big barracudas and groupers nonchalantly cruising past me. Then, I experienced a pang of fear as I considered that I could be a possible prey item for some rogue shark. Later, after I had calmed down, I recalled that the sea is the ancestral cradle of life; I even wondered if I might still retain some vestigial cellular memory of the sea as my ancestral womb. In this vein, consider the experience of eight-year-old Miranda, who sauntered into the ocean one summer day while Mark, her father, looked on:

> [Miranda] stood in the water up to her waist, just moving back and forth with the waves. Ten or fifteen minutes passed, and Mark thought that her eyes were closed. Thirty minutes went by and she was still swaying in the gentle surf in the same spot. After an hour, [Mark] found himself swaying with her as he sat and watched from the beach. It was as if she were in a trance. He wanted to make sure she was all right. Is this some kind of seizure? Does she have enough sunscreen on? he wondered, but he managed not to intrude.
>
> It was nearly an hour and a half before she came out of the water, absolutely glowing and peaceful. She sat down next to him without a word. After a few minutes, he managed to gently ask what she had been doing. "I was the water," she said softly. "The water?" he repeated. "Yeah, it was amazing. I was the water. I love it and it loves me. . . ." They sat quietly until she hopped up to dig in the sand a few minutes later.[16]

Could what Mark witnessed and Miranda experienced be something open to us all? Could it be part of our birthright to know that the sea is our larger body and that our origins are oceanic?

Sylvia Earle (National Geographic explorer in residence) points out: "Even if you never have the chance to see or touch the ocean, the ocean touches you with every breath you take, every drop of water you drink, every bite you consume. . . . Everyone, everywhere is inextricably connected to and utterly dependent upon the existence of the sea."[17] If you are skeptical, consider: What would your life be like without the oceans? Would you still be able to breathe? Not so well—a significant amount of the oxygen you breathe is produced by ocean-dwelling photosynthetic organisms! What about water? Would you have water to drink? Hardly. Without the vast reservoir of water in the oceans—no rainfall, no water cycle, and no benign climate produced by the oceans' ameliorating effects on temperature. Indeed, without ocean currents and associated winds, the tropics would be scorched and the higher latitudes of Earth would be locked into a deep freeze. It is the sea that spreads heat from the tropics to more northerly latitudes, creating the conditions for life to flourish. And what about all the

FISH FOR ANCESTORS!

You have probably heard it said that fish are our ancestors, but, really, how could that be? After all, the idea that some primitive form of fish left the water one day and started walking around on land seems more than a bit fantastical. But, of course, it didn't happen that way. It started gradually, as is often the case with evolution, with fish fins becoming more like forelimbs and with the development of primitive lungs alongside preexisting fish gills. Such features would have been advantageous for ancient fish, enabling them to colonize shallow seas and estuaries. Why? Because these shallow, sometimes stagnant, waters were probably poor in oxygen. The combination of backup lungs and pectoral fins that could also serve as primitive legs allowed some primitive fish species to lift themselves out of water. At first, over millions of years, it is likely that our fish ancestors moved back and forth between sea lagoons and land until eventually they shifted their allegiance entirely to land. The descendants of these land-colonizing creatures gave rise, over vast expanses of time, to such diverse life forms as frogs; dinosaurs; birds; snakes; and most recently to monkeys, apes and, yes, us. As acclaimed science writer Deborah Cramer reminds us: "Though the fish and the sea seem a world apart and though our lives have substantially diverged from theirs, our origins are oceanic. Their lives gave rise to ours, and what they gave us is fundamental to our existence."[18] In a way, I suppose it could even be said that we humans are specialized fish!

grains and vegetables and meat that you eat each day? Without the oceans there to supply rainfall we'd all be out of luck.

As Earle points out: "*Green* issues make headlines these days, but many seem unaware that without the 'blue' there could be no green, no life on Earth and therefore none of the other things that humans value. Water—the blue—is the key to life."[19] After all, it was a nurturing sea that gave rise to Earth's first cell, Earth's first plant, Earth's first animal. Life arose in Earth's seas almost four billion years ago and has continued to diversify and complexify ever since.

The Ocean from Top to Bottom

Imagine being at the beach—say Ocean City, New Jersey—and boarding an expedition boat that takes you hundreds of miles away from shore. There you are, far out in the open ocean, no longer able to see land in any direction. The ocean is vast, seeming to go on forever, and yet you are only seeing the top of it. As you gaze down into the water, it is clear and seemingly empty of life, but looks deceive. If you were to take just one teaspoon of that seawater and examine it under a microscope you would discover a microworld teeming with life—hundreds of thousands of tiny organisms (phytoplankton), all busy using sunlight, water, and carbon dioxide to make food for themselves and the animals of the sea. Although these mostly invisible organisms make up less than 5 percent of all of Earth's plant life, by weight, they account for fully half of Earth's photosynthesis. Translation: The oxygen they generate is part of your every breath.[20]

In that same teaspoon of seawater, there would also be zooplankton (tiny animals, mostly crustaceans, no larger than a *period* on this page) grazing on the phytoplankton. Indeed, just as cows on land feed on grassy meadows, zooplankton in the sea feed on *meadows* of phytoplankton. Think of them as middlemen, converting the teeming masses of phytoplankton into swimming bits of animal protein and fat that serve as food for larger sea creatures.

Though we have names for the Earth's five oceans—Atlantic, Pacific, Indian, Southern, and Arctic—in actuality, these so-called oceans are all interconnected by major surface and deep-water currents that create a single body of water.[21] The global ocean is so vast that only a small fraction (less than 5 percent) has been explored. Indeed, we know more about the moon than we do about the ocean.

Until the last century, humans lacked the means to explore more than a few dozen feet beyond the sea's surface, but these days it is possible to travel to the ocean depths. So, imagine now that you are lucky enough to be entering a modern, high-tech, one-person submersible—a vehicle that can take you to the bottom of the ocean. As you begin your descent, you take in the strange and marvelous underwater world that surrounds you.

Fifty feet: A Portuguese man-of-war drifts by and then suddenly a dense school of herring appear, racing through the surface waters, feeding on both phytoplankton and zooplankton as they go.

One-hundred feet: The sunlight from above catches the water in such a way that you perceive what looks like exceedingly fine *snow* drifting down from above. This *snow* is actually minuscule bits of dead organic matter meandering toward the ocean floor, all the while serving as food for bacteria that, in turn, provide nourishment for deeper-dwelling creatures.

Two hundred feet: You are now well beyond the depth that scuba divers can go. Here you see several comb jellies—semitranslucent creatures with striking bands of iridescent cilia; and then off to your left you spot a solitary Atlantic sailfish.

Five hundred feet: You are now in the "twilight zone." Look up and you detect faint light; look below, total darkness. But here, too, there is life. In fact, the biologists now exploring this zone are discovering new fish species, previously unknown to science, almost everywhere they look.

One thousand feet: You are beginning to enter the ocean depths—Earth's largest habitat. The water gets colder and colder as you descend and the pressure ever more crushing. The animals down here are tough; the small ones survive by foraging on the rain of organic debris from above; the larger ones by eating the smaller ones and each other. This is the home of the giant squid (*Architeuthis dux*), a creature with eyes as big as Frisbees and a body as long as a school bus.

Now, looking into the abyss below, you are startled to see flashes of light here and there. At times it is almost as stunning as a glimpse of Las Vegas at night. In those inky depths live hundreds of species of fish and invertebrates that have the ability to create light using chemicals in their bodies. Why light up? Sometimes their flashes help with navigation; sometimes they act as a warning to back off; sometimes they are a mating signal; sometimes a trap. For example, the anglerfish has a lighted *lure* extending from the top of its head that draws fish close enough so that the angler can snatch them up.

Ten thousand feet: Now, approaching the seafloor, you see, rooted in the substrate, what appear to be clusters of shrubs, six feet tall, topped with red plumes. These are giant tubeworms—creatures lacking both mouth and digestive system that survive by using their red plumes to take up oxygen from the seawater along with hydrogen sulfide emitted from hydrothermal vents on the ocean floor. These gases serve as substrates for bacteria that live in the tubeworm's body. The bacteria, through a process known as chemosynthesis, oxidize the hydrogen sulfide and in the process create energy (organic food molecules) that the tubeworms need to survive. The bacteria, for their part, get a nice place to live and may comprise up to half the weight of a tubeworm's body. As you inspect this bizarre seabed environment more closely, you note the presence of specially adapted species of fish, crabs, shrimp, starfish, anemones, among other creatures. Yes, ocean life goes deep.[22]

Ocean sediment: Your little sub comes to a stop on the ocean floor. If you were able to dig into the sedimentary layers of sand, silt, and shells extending thousands of feet below you, you'd find life there too—lots of it! Indeed, scientists drilling into the seabed typically find a billion or more organisms in just a half-teaspoon of sediment. And these organisms are utterly new to science, distinct in their lineage from all other microbe groupings. But, in spite of their extraordinary abundance, they are just barely alive. For example, their metabolism is so slow that they live in a kind of suspended animation, dividing (i.e., reproducing), perhaps, only once every thousand years.[23]

Before you ascend, you take one last look around and something strangely out of place catches your eye. You maneuver your sub closer and are surprised to discover an empty Coke can. You explore a bit more and find another beverage container. Though very few humans will ever get to go to the bottom of the ocean, our stuff is now finding its way there. For example, my colleague Charles Fisher has taken more than a hundred trips in submersibles and reports that on each trip, almost without exception, he has encountered some trash on the ocean floor.

Checking the Pulse of the Sea

Given the immensity of Earth's oceans it seemed unlikely, at least until recently, that humans could affect this vast biome. But just in the past hundred years, humans have caused more harm to Earth's oceans than during all preceding history and the pace is quickening. Indeed, that Coke can on the ocean floor appears to be just the tip of the proverbial iceberg.

Ocean Fisheries Sinking

It was in the late 1980s that the worldwide ocean fish catch (i.e., the total weight of fish caught per year) peaked at about 80 million tons. Since then, in spite of new, more efficient fishing technologies, the catch rate is no longer rising. More worrisome still is the fact that the composition of the ocean fish catch is changing. The biomass of large predatory fish like tuna, marlin, swordfish, cod, halibut, and grouper has declined by as much as 90 percent compared to a century ago.[24]

Much of the decline in the catch of large ocean fishes has occurred since the 1950s with the introduction of highly aggressive fishing technologies. For example, some fishing vessels set out "longlines"—fishing lines that extend for dozens of miles, with as many as three thousand baited hooks distributed along each line. Meanwhile, industrial trawlers act like bulldozers scraping along the ocean floor, scooping up everything in their path, including the seabed habitat itself. This results in the inadvertent capture of many creatures that are not of commercial interest. The stuff not wanted is called "bycatch." These days, for every pound of fish that goes to market, ten or more pounds are thrown away as bycatch. It's even worse in the case of shrimp: For every bushel of shrimp harvested, a hundred bushels of sea life are pulled on board in driftnets.[25] Nor do marine mammals, sea birds, and turtles escape harm. Data assembled by the World Wildlife Fund reveal that upward of 300,000 marine mammals and hundreds of thousands of turtles and seabirds are inadvertently killed each year in industrial fishing operations.

As aggressive fishing operations exhaust the stocks of large commercial species, smaller fish like anchovy, shad, and mullet—that the large preda-tory fish would normally consume—have become a new target for com-mercial fishing. And when these smaller fish are gone it is possible that the whole ocean system could lose its integrity. What does that mean? In the words of biologist Tierney Thys, "With fewer fish, low-energy animals such as jellies can bloom in great numbers and gain a stronghold on the sea."[26] Thys's words evoke the airplane-rivet-popping illustration from earlier in this chapter: the more we simplify the ocean community by removing its rivets (i.e., jeopardizing its species), the more we undermine the ocean's complex web of relationships. The result is an "aircraft" (think planet Earth) that is less safe and more prone to disasters.

Dead Zones Expanding

In the early 1970s biologists who were surveying the waters of the Gulf of Mexico off the coast of Louisiana were mystified when they discovered that seawater near the bottom had a milky appearance. When the researchers analyzed these bottom waters, they found they had elevated concentrations of nitrogen and phosphorus and were almost devoid of oxygen, and, there-fore, life. As they continued monitoring in subsequent years, they noted that the Gulf dead zone expanded each summer and then contracted in the fall and winter.

Put yourself in the place of those researchers endeavoring to solve this mystery. Why might those bottom waters be enriched in nitrogen and phos-phorus? And why might they be almost devoid of oxygen? As you ponder this question, recall what you learned in chapter 2 about how disturbances such as deforestation interfere with natural cycles, causing nutrients (e.g., nitrogen, calcium, phosphorus), to leak from the land. Now picture the massive amounts of water flowing out of the Mississippi—more than a mil-lion gallons every second—into the Gulf. This water originates as rain fall-ing on the thirty-one farm states that constitute the enormous Mississippi basin watershed. Before the advent of agriculture, the waters of the Missis-sippi would have run clear and clean (except during floods); but these days, with the land cleared for farming and the annual application of millions of tons of fertilizers, enormous amounts of nutrients leak into the Mississippi each year. Eventually these nutrients are carried to the Gulf where they stimulate a summer surge of algal growth. Following this growth pulse, the short-lived algae die and settle to the seafloor where they serve as food for microbes. Inundated with food, the microbe populations explode and, because they need oxygen to survive, they quickly exhaust the oxygen supply in the bottom waters. This results in the death of many bottom-dwelling organisms—including oysters, mussels, and clams—and, ultimately, in the creation of dead zones.[27]

The Gulf dead zone has been increasing since the 1970s and now, in most years, covers more than five thousand square miles. And this is not a phenomenon exclusive to the Gulf; there are now more than four hundred dead zones worldwide. Indeed, expanding dead zones have been identified virtually everywhere on Earth where major rivers flow into the ocean.[28]

Plastic Oceans

If you have ever gone beachcombing, chances are you have encountered bits of trash among the shells and driftwood. That's to be expected insofar as we humans are sometimes careless with our trash. What would not be expected—at least I never imagined it—is that enormous patches of human-generated garbage now exist far out in the open seas. The most famous such patch, known as the "Great Pacific Garbage Patch," is located in the North Pacific Gyre—a zone of seawater circulation that draws in trash over an area twice the size of Texas. In 1999 Charles Moore pulled a fine mesh net through this ocean region and then sorted through the contents comparing the weight of living organisms (mainly plankton) to that of plastic. The result: Life lost 6:1—that is, for every ton of living creatures there were six tons of rubbish.[29] The North Pacific Gyre is one of six such ocean zones; there are others in the South Pacific, the North and South Atlantic, and the Indian Ocean. Combined, these garbage-collecting gyres represent 40 percent of the sea, with plastics making up 90 percent of their garbage.[30]

This plastic debris won't go away anytime soon. Indeed, except for the tiny amount that's been incinerated, every molecule of plastic ever made is still around. Fifty years ago we weren't making much plastic, but nowadays industries turn out in excess of sixty billion tons of the stuff a year, much of it intentionally designed to be used one time only. No surprise, then, that each American disposes, on average, almost two-hundred pounds of plastic each year.[31]

Given enough time, sunlight will cause plastic to photo-degrade, reducing it to smaller and smaller particles, but never wholly eliminating it. All this accumulating plastic has direct and deleterious effects on sea life, in part because many plastic pieces look like food to sea animals. Consider:

- In 2007, a whale found dead off the coast of California had four hundred pounds of plastic in its stomach.
- The stomachs of dead seabirds are sometimes filled with the likes of colored plastic bottle caps, plastic foam, toy soldiers, and Lego parts.
- Nearly all fish examined in the Central Pacific by researchers have at least some plastic fragments in their guts.
- The same is true for small filter-feeding organisms like clams, oysters, and mussels, as well as large plankton-feeding creatures such as manta rays and whale sharks.

"Whether at the large, medium, small, or ultra-small scale, ingested plastic . . . kills by physically obstructing, choking, clogging, or otherwise stopping up the passage of food."[32]

And it is not just physical obstruction that is cause for concern; it is the actual chemical nature of plastics. True, some are inert and harmless, but many are imbued with additional chemicals (e.g., dyes, flame retardants, softeners) and none were intended for consumption by marine creatures. Moreover, the open molecular structure of plastic fragments renders them sponges for the absorption of toxins like mercury, PCBs (polychlorinated biphenyl), and pesticides that are present in the ocean in dilute quantities. For example, chemists examining plastic pellets found in waters near Japan reported levels of DDE and PCBs (both probable carcinogens and neuro-toxins) a million times higher than in the surrounding seawater. One only need to consider eating a plate of oysters laced with toxin-laden minuscule plastic particles to imagine how a "plastic-infused ocean" could come home to roost in our very bellies. This serves as yet one more reminder of how our lives are inextricably bonded to the life of the sea.[33]

STEPPING BACK TO *FEEL* THE BIG PICTURE

The depth of our feeling life measures the depth of our life force, and if we judge, contain, or repress our feelings we repress our life force.

—Ingrid Bacci[34]

Having just learned how the populations of many bird species are in decline, are you able to imagine what it would be like to actually be a bird that tries to reproduce only to have her habitat destroyed or her freshly laid eggs eaten by a feral cat? Or can you imagine what a whale might think, as well as feel, when it encounters—as surely it must—the Great Pacific Garbage Patch?

As we begin to learn about, and open our minds and hearts to the battered condition of our home, planet Earth, some common responses are:

- Why didn't anyone ever tell me about this before?
- I want to scream (or cry) when I think about what humans are doing to Earth.
- I care about Earth and what we are doing to it, but no one else seems to care.

Actually, hundreds of millions of people around the world share these concerns and sensitivities, but most remain silent, reluctant to voice their views publicly.

Ecopsychologist and activist Joanna Macy believes that it is fear that is at the root of this silence. First, according to Macy, we fear the pain and guilt that we might experience if we were to risk opening our minds and hearts to the multitude of ecological problems now plaguing Earth. Second, we fear the humiliation and shame we might experience if we do speak out, but lack all the necessary facts to defend our views. Finally, we fear that if we speak our true feelings and concerns about the plight of Earth, others will judge us odd, overly emotional, or impractical.[35]

As a counterbalance to these fears, Macy emphasizes that the despair and pain that many people feel concerning the damaged state of Earth stems from their compassion; it is a measure of their humanity and their sense of connectedness to Earth. "Pain is the price of consciousness in a threatened and suffering world," writes Macy. Feeling pain "is not only natural, it is an absolutely necessary component of our collective healing. As in all organisms, pain has a purpose: it is a warning signal, designed to trigger remedial actions."[36]

We can, I believe, begin to free ourselves from the bondage of fear and despair by simply paying attention to the feelings and emotions that arise in us when we hear about different aspects of the ecological crisis now gripping Earth. What arises in you, for example, when you read about dead zones in the ocean, decimated fisheries, plastic-infested seas, fragmented forests, dying birds? Specifically, what happens in your body—in your jaw, neck, shoulders, eyes? What happens to your breathing? And what about your thoughts? Is there judgment, resistance, anger, fear? Do you want to run away? Do you feel helpless? Numb? Indifferent?

Sometimes, in helping people to make contact with and to express their capped-over feelings regarding the condition of Earth, I select objects that symbolize different emotional states and place them in the middle of our gathering space. For example, if the destruction of Earth causes your heart to feel cold and numb, you could pick up a cold, hard rock and, holding it, speak of the numbness you feel. If it is anger that you feel, you might pick up a strong stick and, holding it fiercely in both hands, give voice to your anger. If you are feeling sadness, you could pick up dead leaves and speak of your aching heart as you clutch those withered leaves. In a like vein, you might access feelings of emptiness, confusion, or hopelessness by cradling an empty bowl in your lap.[37]

The purpose here is not to inflict new pain. Rather, it is to bring to consciousness the capped pain that I suspect we all carry to varying degrees. We carry this pain because we are, inescapably, an expression of the living—and in these times wounded—body of Earth. As psychologist Ingrid Bacci points out, "Our world is in a mess right now because too many of us have lost touch with the art of caring. We need to learn how to care, and there is only one way to begin doing that—to practice experiencing our feelings."[38]

The Vietnamese monk Thich Nhat Hanh has said that to heal the world the first thing we need to do is "hear, within ourselves, the sounds of Earth's crying." We can only do this, I believe, to the extent that we are in relationship with Earth. For example, when I learned, some years back, that an interstate highway was going to be placed right down the middle of the beautiful valley where I live in central Pennsylvania, I felt moved to respond in some way. The route had been flagged; soon the explosives and the heavy machinery would arrive. Nothing could be done, I was told, but this was not true. I couldn't stop the road, but I could still bear witness to the loss of the meadows and forests and farms that this road would destroy. So I resolved to simply walk the land that was destined to become Interstate. In so doing I hoped to experience this land, thank it, and say good-bye to it. The idea of a good-bye walk resonated with others in my community. So it was that a group of some fifty citizens gathered on a cold Sunday in February to walk that swath of land that would soon be entombed in concrete. We went in silence, stopping here and there along the way to reflect. Our walk ended late in the afternoon on the banks of Spring Creek. The silence continued, punctuated from time to time by expressions of grief, tenderness, gratitude, sorrow, song. In retrospect, I now see that on that day we were all learning to "hear, within ourselves, the sounds of Earth's crying."

TO FEEL IS OUR BIRTHRIGHT

As a male, I was brought up to regard my feelings as something to keep under cover or, at the very least, to keep under control. In school, when I fidgeted around in my seat—something I apparently did frequently—I was told to get myself under control. When I fell and skinned my knee in the schoolyard, I was warned, "Big boys don't cry!" It's as if there was a sign at the entrance of my school saying, "Check your emotions at the door!"

I was just a kid and I wanted to please my teachers; and in my kid's head I assumed that my teachers must be right. After all, they'd been around a lot longer than I had. So, if they said that expressing exuberance or anger or sadness or fear was unacceptable, who was I to argue!

Looking back, I think I understand what was going on. Emotions are energy. So, in school, when I felt my emotions, they moved me; I became enlivened and my actions were based on an inner authority. But the various schools I attended were, for the most part, designed to make me outer directed—that is, they were structured to channel my attention and allegiance toward outer authorities—teachers, institutional rules, textbooks. Thus I, along with most young people still today, grew up spending tens of thousands of hours seated at a desk, learning to face forward, pay attention, follow instructions, and, through it all, to stuff my emotions and feelings deep inside. Only now am I beginning to wonder about the unacknowledged toll this might take on our individual and collective spontaneity, creativity, exuberance, authenticity, and more.[39]

WRAP UP: EXPANDING ECOLOGICAL CONSCIOUSNESS

No other quality is so urgently needed today [as listening]: Millions cry out to be heard, to be listened to, to be allowed some say in a harsh and brutal world. Planet Earth itself cries out amid its pain of pollution, exploitation and desecration.

—Diarmuid O'Murchu[40]

Environmental scientists, like vigilant doctors, have been *listening* to the creatures of the land and sky and sea and the messages we have been receiving reveal that all is not well on planet Earth. Yes, we humans are fomenting environmental havoc. But this need not be. We are more than mindless automatons; we are bearers of reflective consciousness. Our very name—*Homo sapiens sapiens*—the one who knows that s/he knows—bears testimony to our capacity for reflection.

These times, I believe, are calling us toward a fundamental shift in consciousness—in how we see ourselves in the context of the larger community of life. At present, human actions in the world are most often grounded in anthropocentrism—the belief that humans are top dog and that Earth is here for us to use as we see fit.[41] This worldview has provided us with a convenient justification for the elimination of plant and animal populations from areas that we covet, whether on land or in the sea. Even in those instances where we have taken steps to protect wild species, our rationale is often imbued with anthropocentrism. For example, we might decide to protect certain tropical ecosystems, believing that their resident plant and animal species may contain compounds or genes that will help cure human diseases or enhance our crop-breeding programs. Even when we set aside parklands, we do it primarily for our own enjoyment—to serve *our* recreational ends.

Ecocentricism—valuing and respecting all species and ecosystems to the same degree that we value and respect ourselves—stands as a firm counterpoint to anthropocentrism. An ecocentric orientation would lead humans to embrace the *Noah Principle* that states—plain and simple—a species' long-standing existence on Earth carries with it unimpeachable rights to continued existence. In other words, the simple fact that a species has come into being and is now a living part of Earth's web confers on that species the right to continued existence. Although humans have not yet adopted the Noah Principle, we have taken some important steps. For example, over its short history, the United States has extended rights and acknowledged moral obligations to disenfranchised segments of society—for example, to children, women, people of color. Likewise, through legislation such as the Endangered Species Act, the United States has begun to consider its moral obligations to other species.

Ideally, this progression toward an ever wider and more inclusive sense of moral obligation develops over the course of a lifetime (see figure 4.2).[42] As young children, our sphere of awareness and concern is limited to immediate family members. Then, in grade school, awareness typically expands to friends and neighborhood residents; later, in high school, the purview of concern extends to the local community; and by college, if all goes well, we begin to experience ourselves as citizens of a nation. With further maturity, ideally, we come to see ourselves as integral part of planet Earth, as well as the cosmos. When this happens for humankind as a whole, we will be living in ecocentric consciousness. Getting there isn't easy, but it is possible, provided we have the courage to grow up and assume our birthright as generative, fully conscious beings.

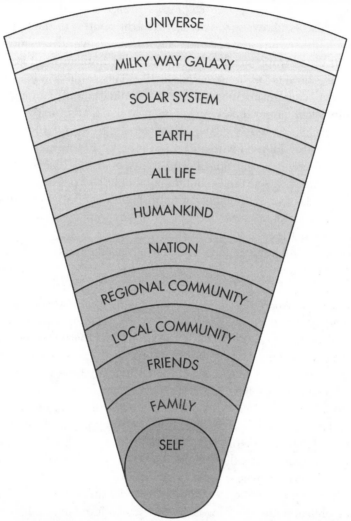

Figure 4.2. The expansion of human awareness from child-hood through late adulthood.

APPLICATIONS AND PRACTICES: LISTENING

> We have two ears and one mouth so that we can listen twice as much
> as we speak.
>
> —Epictetus

Given the difficulty many of us encounter in our attempts to listen to each other, is it even possible to become attuned to the lives of other creatures?

Even the idea of extending our attention to wild creatures may seem crazy, especially if we are stuck in the anthropocentric consciousness of species-ism. But for someone with an ecocentric perspective, the squirrel in the backyard is not *just a squirrel*, it is a unique, one of a kind individual.

Barbara Smuts, a primatologist at the University of Michigan, learned to see animals as individuals when she lived with a baboon troop in the wilds of Kenya. Eventually, Smuts was able to identify all one hundred fifty individuals in the troop based on appearance and behavior. One day, after she could indentify everyone, one of the dominant males in the troop looked in her direction and then flattened his ears against his skull while simultaneously raising his eyebrows so that the white skin on his eyelids was visible. Smuts knew that this was a signal used by males to beckon females to them. So she looked behind her to see which female baboon this male was signaling, but there were no baboons behind her. This left Smuts both astonished and delighted. Keeping her distance, she behaved in a baboon-appropriate way to acknowledge the male's presence and to signal respect. This story is remarkable insofar as it reveals that it was only when Smuts was finally able to see the baboons as individuals that they reciprocated by seeing her—no longer as an object—but as a subject with whom they could communicate.[43] Reflecting on her time with baboons in Africa, Smuts writes:

> Before Africa, if I were walking in the woods and came across a squirrel, I would enjoy its presence, but I would experience it as a member of a class, "squirrel." Now, I experience every squirrel I encounter as a small, fuzzy-tailed, person-like creature. Even though I usually don't know one squirrel from another, I know that if I tried, I would, and that once I did, this squirrel would reveal itself as an utterly unique being, different in temperament and behavior from every other squirrel in the world. In addition, I am aware that if this squirrel had a chance to get to know me, he or she might relate to me differently than to any other person in the world. My awareness of the individuality of all beings and of the capacity of at least some beings to respond to the individuality in me, transforms the world into a universe replete with opportunities to develop personal relationships of all kinds.[44]

Listening to the More-Than-Human World

There may be no better way to cultivate ecocentrism than to learn to pay attention to the wild creatures that we live among. Here is a simple starter practice: The next time you are in a natural setting, allow yourself to wander about, simply paying attention to the things that you find pleasing. Think of it as an adventure. The idea is to simply go where your feet take you.

As you explore, know that the beings nearby—though you might not see them or hear them—are aware of your presence—attuned to your movements. Know, too, that it is possible to develop a relationship with these denizens of the wild. All that is required is that you walk slowly, without any quick or jerky movements, and that you practice humility by conveying friendly intentions with your voice, your gestures, your facial expression and your body posture. This will reassure creatures close by that it is safe to go about their daily activities.[45]

Eventually, a time will come when something in particular attracts you. It might be as seemingly mundane as a millipede or a beetle or a bee or a pill bug or a rotting stump. Whatever it is, ask its permission to spend time with it. If it strikes you as odd to ask permission, consider how you would feel if someone came barging into your house and generally swaggered around as if s/he owned the place. Contrast this to how you would feel if that person politely knocked and asked if s/he might visit with you. Viewed from this perspective, when you walk into a forest, you are literally walking into the *home* of other beings. So asking permission is simply a way of extending the same respect to the wild dwellers of the forest that you, presumably, would want others to extend to you.

When you are ready, take time to explore this subject that has garnered your attention. You have not arrived randomly; you are there so that you might learn something from this *Other*. Notice how the subject of your attention is simply itself—utterly real, authentic, with no pretense. Open your senses and trust that some deeper understanding of who you are in relationship to this *Other* will emerge. This is a time for you to look both outward and inward—*listen*.

As you relax, consider the connections—the web strings—between this *Other* and your own inner depths. After all, you are related to—*a part of*—everything that you see and touch and smell and hear and taste, just as every being around you is a part of you, whether you know it or not. Then, when the time comes to end your encounter, take a moment to simply express gratitude in whatever way seems appropriate.

In sum, to become more intimate with the life around you, it is helpful to cultivate a spirit of innocence, openness, and curiosity—what some call "beginner's mind." Rather than seeing the bush, the ant, the bluebird, the red spruce through the veil of your fixed thoughts and opinions, you can endeavor to see them with an uncluttered mind and fresh eyes.

Listening to the Human Other

These days, when suffering is everywhere—in the wars that divide peoples, in the greedy exploitation that defiles ecosystems, in the psychic depression that shrivels hearts—we have, arguably, no more important task than to learn how to listen to each other. The simple act of listening with an open heart, free of judgment, invites spaciousness and generosity into the world.

Listening can be especially challenging when we are hearing things that we disagree with. For example, imagine the following situation: You are sitting in a coffee shop when the guy at the next table throws down his newspaper and exclaims, "Enviro-tree-huggers are frigging nuts. They are trying to stop everyone from making a living."

You look over and catch his eye, signaling a willingness to listen. This is not going to be easy because you are something of an enviro-tree-hugger yourself. You know arguing won't work, so you decide to try something new. Instead of getting hung up on the man's words, you resolve to place your attention on the feelings under his name-calling. So it is that you are able to nod empathetically and say, "You're pissed off about the enviros?"

He shoots back, "They are a bunch of spoiled brats, and they don't know jack."

"They really piss you off," you reply.

"Damn right, I'm pissed. My brother and uncle are both loggers, and their jobs are on the line if this stuff keeps up."

You pause; and then once again, focusing on what you hear as the man's feelings, you say, "You're worried about how the stuff these enviros are doing could affect your people."

"Yeah, ya know it's not easy being a logger these days," he responds.

"And you're worried about your brother and uncle," you offer.

"Well, yeah. Ya know, they're family."

You sense a softening in the man. As you listen without judgment, you are able to go beyond his initial complaint about the "friggin' enviro-tree-huggers" (which was actually a judgment born of the pain the man was feeling) to his underlying emotions. It turned out that he was feeling a combination of anger, worry, and care. When you were able to reflect his underlying feelings back to him, he felt heard.

When we listen to each other in this way, we can detect the unmet needs that are always lurking below harsh judgments. Marshall Rosenberg, in his book *Nonviolent Communication*, suggests a practice that all of us can use to become more aware of the connections between our own judgments and needs:

> List the judgments that float most frequently in your head by using the cue, "I don't like people who are _____." Collect all such negative judgments in your head and then ask yourself, "When I make that judgment of a person, what am I needing and not getting?" In this way, you train yourself to frame

your thinking in terms of [your] unmet needs rather than in terms of judgments of other people.[46]

For example, if you are in a conversation with a man who keeps talking
at you a mile a minute, you may judge him as self-centered and boring.
But what if you follow Rosenberg's suggestion, and rather than sitting in
judgment, you simply ask yourself, "What am I needing and not getting in
this moment?" In this case, you might discover that what you need—what
you are really yearning for—is a conversation that engages you by inviting
your participation. With this realization you can, if you choose, shift from
silent judgment (which leaves you feeling separate and victimized) to the
active expression of your needs and, in so doing, create the possibility for
genuine relationship.

Once we understand that our judgments are invariably grounded in unmet needs, we may find it easier to think of ways of resolving differences.
For example, continuing the imaginary conversation with the man in the
coffee shop, you could move from feelings to needs by saying, "Sounds like
it is important for you to know that your brother and uncle aren't going to
be out on the street without a job."

To this the man might respond: "Yea, they both have young kids. Without
a job they'd be screwed."

As you continue to talk, it becomes clear that this man doesn't harbor any
ill will toward environmentalists per se. In fact, he confesses that he likes to
fish and that all the runoff and erosion associated with forest clear-cutting
is messing up one of his favorite fishing streams. Ultimately, this man's
fundamental need is for the safety and well-being of his family. In this he
shares immense common ground with the so-called enviros.

This scenario validates the old adage: If you hate someone or condemn
someone, you just haven't heard their story yet. Indeed, listening without
judgment creates enormous space for understanding and reconciliation.

Rather than merely reading about this encounter between the logger
and the enviro, we can, if we choose, apply it to our own lives. If you are
game, you might begin by identifying someone in your life who is causing
you stress—it could be a friend, a parent, a brother or sister, a boss. Next,
call to mind all the things about this person that are upsetting to you and
then write down what you think of them. Include what is it about them that
angers, disappoints, or confuses you, and why. Let yourself really rant away.
Finish by describing how you want them to change.

When you are done, read over what you have written and consider how
some of your judgments about this person might actually not be true. After

all, judgments are just beliefs and beliefs are just opinions. So see if you can soften a bit. In a similar vein, open yourself to considering that some of your judgments about this person might actually apply as much, if not more, to you as to them. For example, if you are upset with a housemate because s/he ignores you, see if you can find instances where you ignore others, including your housemate. Again, the idea is to see if you can soften a bit. This softening is not easy because our egos are hugely invested in being right and they love to blame and judge.

Now for the really challenging part: Make arrangements to spend an hour with this person who is causing you stress with the sole intention of understanding them—free of judgment, free of the need to be right, free of your ego's agenda. Your only purpose during this hour is to listen to and seek to understand this person. You are *not* interrogating. Rather, you are just inquiring about their life, their concerns, their struggles, their joys. Your sole intention is to see and understand the world through their eyes. If the person says something that you disagree with, avoid the inclination to enter into a debate, and instead simply say something like, "I am confused by what you are saying; please help me understand you." In carrying out this project you will be, in effect, testing a hypothesis—namely that *it's impossible to be curious about someone and, at the same time, to stand in judgment of them.*[47]

QUESTIONS FOR REFLECTION

- What, if anything, would you do if you saw a cat stalking a migratory songbird? Why?
- Where did you feel most at home in nature as a child? What has happened or is happening to your favorite childhood haunts?
- What feelings come up for you when you think about the Great Pacific Garbage Patch? Where in your body do you have those feelings?
- In what ways have you been silenced by fear (a la Joanna Macy) and what, if anything, might you do about this?
- What would it mean for you to hear, within yourself, the sounds of the Earth's crying?
- What are your most common judgments of others and what personal unmet needs lie below these judgments?

5

COURAGE

Facing Up to the Unraveling of the Biosphere

It is obvious that something is not working properly. How else can we explain our mismanagement of resources and of our own population, the pollution and the destruction of our environment, or the mass murder of our own species? We cannot see the bigger picture and how we fit into it. We are no longer in our bodies; we are not in our right minds.

—Wes Nisker[1]

The title of this chapter has an unnerving quality. Could it be that our home, planet Earth, is really unraveling? Could it be that today Earth is profoundly different—irrevocably changed—from the Earth that existed when our great grandparents were born? I don't mean *different* because we now have all kinds of nifty technologies that our grandparents lacked. I mean *different* from the perspective of Earth—*different* biologically, functionally. Here's an analogy. When you become sick, if you are like most people you behave differently. For example, you don't move the same way, don't eat the same way, don't act the same way; you are off-kilter. After a time, though, your body heals itself and soon, with any luck, you are back to your old self. But imagine that you didn't recover; that you remained mired in sickness for the rest of your life—never again to return to your healthy self. Could it be that this is what is now happening to Earth? Might it be that we humans are *perilously close* to pushing Earth beyond its capacity for self-regulation—that is, beyond its ability to self-heal on any time scale relevant to our continued survival? Indeed, an increasing number of scientists fear that we may *have already gone too far*—that we now live on a feverishly sick planet that will become increasingly inhospitable to humans, including you and me, in the coming decades.

I tell you up front that I write with fingers crossed, desperate to believe that we have not, already, gone too far. I don't allow myself to drop into the doom-and-gloom camp because believing that we are doomed engenders numbness, apathy, and deadness in me. In truth, none of us—not even those certified climate scientists with their complex models and prognostications—knows for sure what lies ahead.

In this chapter, I invite your courage in looking fearlessly at our prospects, in light of such global phenomena as climate change, thinning soils, depleted aquifers, and chemical contamination of the air we breathe, the water we drink, and the food we eat. Admittedly, none of this is very cheery, but our future prospects, I believe, will improve to the extent that we pay attention, stay awake, and avoid falling into the potentially lethal trap of denial.

FOUNDATION 5.1: CLIMATE OUT OF CONTROL?

Today, our prevailing consciousness leads us to simply see air as empty space or, if you are of a scientific disposition, to quantify the atmosphere in terms of its gas mix. This purely mechanistic way of seeing fails to acknowledge that each of us is intimately entangled with Earth's life-giving atmosphere. The surfaces that comprise our skin, lungs, internal organs, tissues, and muscles are bathed with atmosphere and inseparable from it. Whether we know it or not, we are in relationship with Earth's *airscape*, and what we do to it, we, necessarily, do to ourselves.

Our Changing Airscape

Earth's airscape is dynamic, as evidenced in figure 5.1.[2] This is arguably the most important graph produced in the history of environmental science. The figure reveals a steady upward trend in the concentration of carbon dioxide in Earth's atmosphere since the 1950s.

A fascinating aspect of this graph is the annual oscillation in atmospheric carbon dioxide concentrations—with a rise and a drop each year. These oscillations mirror, in a sense, Earth's *breathing*. To understand how this is so, imagine that you were able to create a self-contained forest microcosm

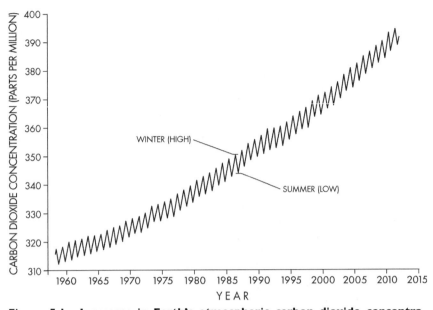

Figure 5.1. Increases in Earth's atmospheric carbon dioxide concentrations since 1958.

with its own private atmosphere. Now, picture this forest in winter. The leafless trees have no means of gaining energy through photosynthesis and so, to stay alive they must break down sugars stored in their stems and roots, in the process releasing carbon dioxide (CO_2) to the microcosm. All the animal life in this forest microcosm will also continue to respire, breathing out CO_2. Hence, carbon dioxide concentrations in the atmosphere of this forest microcosm will rise during the winter. Now, fast-forward to May: Trees are now flushing out new leaves. Since carbon dioxide is a foodstuff for photosynthesis, these new leaves begin to remove carbon dioxide from the atmosphere. Of course the trees and animals will continue to release some carbon dioxide to the atmosphere via respiration, but the carbon dioxide uptake by the photosynthesizing trees will be considerably greater than the carbon dioxide release. Hence, carbon dioxide concentrations in the atmosphere of the microcosm go down from late spring through early fall. This, in effect, is what is happening on a global scale.[3] Because there is more land and more trees in the Northern Hemisphere, Earth *breathes in*—that is, removes carbon dioxide from the atmosphere) during the Northern Hemisphere spring and summer (creating the dips in figure 5.1), and *breathes out* carbon dioxide during the Northern Hemisphere fall and winter (creating the peaks).

There is another point to be garnered from this graph—namely that carbon has its own cycle. Just like calcium and other elements (see chapter 2), carbon moves among various storage places, or pools. The main pools for carbon include forests, soils, oceans, and the atmosphere. For almost all of human history these carbon pools have been fairly stable, but then, starting about two hundred years ago, humans began to tap into enormous new pools of carbon—especially oil, coal, and natural gas. Burning these fossil fuels for energy has resulted in the utter transformation of human life and the environment. Here's why:

> One barrel of oil yields as much energy as twenty-five thousand hours of human manual labor—more than a decade of human labor per barrel. The average American uses twenty-five barrels each year, which is like finding three hundred years of free labor annually. And that's just the oil; there's coal and gas too. . . . [Fossil fuels are why] we're prosperous, why our economies have grown. It's also, of course, why we have global warming and acid oceans; in essence we've spent two hundred years digging up all that ancient carbon, combining it with oxygen for a moment to explode the pistons that take us to the drive-through, and then releasing it into the atmosphere, where it accumulates as carbon dioxide. . . . All day every day we burn coal and gas and oil, from the second we make the coffee till the second we turn out the lights. . . . If an alien landed in the United States on some voyage of exploration, he might well report back to headquarters that we were bipedal devices for combusting fossil fuel.[4]

Learning to mine and burn fossil fuels has been a tremendous boon to humankind. However, nowadays, the relentless combustion of fossil fuels is causing Earth's climate to warm with potentially disastrous consequences. This should come as no surprise. After all, you've seen time and again in this book how everything is connected—how it is impossible to do one thing without creating ripple effects.

What Exactly Is Global Warming?

The physics of global warming is grounded in two simple concepts:

- Energy enters Earth as sunlight and leaves in the form of infrared radiation (heat). For temperature to remain constant, the energy reaching Earth must approximate the energy leaving Earth.
- Greenhouse gases, like carbon dioxide, absorb infrared radiation (heat) but not sunlight. Hence, when the concentration of greenhouse gases increases, some of the infrared heat energy, which previously escaped to space, is intercepted and reflected back, warming Earth's surface.

HOW COULD THIS BE TRUE?

It seems like an outrageous proposition to suggest that we, mere humans, could actually change the climate of something as grand as planet Earth. But Earth's atmosphere doesn't extend endlessly out into space; it is only a wafer-thin protective layer. How thin? Imagine Earth scaled down to the size of a basketball. Dip this ball in water and pull it out. That gossamer-thin sheen of water adhering to the ball represents, to scale, the extent of Earth's atmosphere. Human activities can and do affect this narrow, life-sustaining band.

Some trapping of infrared radiation is a good thing. Without it, Earth would be brutally cold—with the oceans frozen from top to bottom. At the other extreme, if none of the solar radiation entering Earth's atmosphere was able to escape as heat, Earth would quickly heat up to unbearable temperatures. Maintaining the right balance is important. Humans have upset this balance primarily by combusting carbon that was once locked away deep in the ground and releasing it into the atmosphere, where it acts like a heat-trapping blanket.

What Does Climate Research Reveal?

If your great-great-grandfather had climbed to the top of a remote mountain in 1850 and collected a bottle of air; and, if you were to go to that same remote mountain today and collect a new sample, the two air samples would be substantially different. Their basic chemistry would have changed. We need to let this fact register. "The air around us, even where it is clean, and smells like spring, and is filled with birds, is different, significantly changed."[5]

The difference has to do with the gas makeup of the atmosphere. Gases in the atmosphere are typically measured in terms of parts per million (ppm).[6] In 1850 Earth's atmosphere contained about 280 parts carbon dioxide per million, but this was before humans started burning oil, coal, and gas in earnest. At the time of this writing the CO_2 concentration of Earth's atmosphere is topping 400 parts per million; and this is well above what scientists consider a *safe* upper limit of 350 ppm. Indeed, the greater the CO_2 increase beyond 350 ppm, the more climate-related disruption we humans are likely to be subjected to.

Not only has the amount of carbon dioxide been going up steadily in recent history, but the rate of increase has also been rising, decade by decade. For example, in the ten-year period, 1992 to 2001, the average annual rise

BUT WHAT ABOUT THE NEWS REPORTS DENYING CLIMATE CHANGE?

A vanishingly small number of scientists contend that human activities are in no way responsible for the current warming trends on Earth. These dissenting voices sometimes, though not always, receive their funding (directly or indirectly) from the fossil-fuel conglomerates that have a vested interest in making the public believe that significant disagreement exists among scientists regarding climate change. Meanwhile, media outlets—often on the lookout for controversy because discord attracts readers and viewers—sometimes bend over backward to give significant airtime and print space to climate-change naysayers, creating the impression that the scientific community is deeply divided. None of this is to say that there are not disagreements among scientists regarding climate change. Indeed, there is healthy argument about the long-term impacts of climate change and how we should respond, but there is virtually no disagreement among scientists about whether it is happening.[7]

in atmospheric CO_2 concentration was 1.6 ppm; in the following decade (2002–2011) the average annual rise increased to 2.1 ppm. Given current trends, atmospheric carbon dioxide concentrations will be in the vicinity of 500 ppm by 2050. If this happens, the vast majority of climate scientists believe that all human life-support systems on Earth will be in significant disarray as we are confronted by a world warmer than anything humankind has ever known.

It takes courage to question what we've encountered in the popular press or overheard on the street, and draw our own, evidence-based conclusions about climate change. The logical starting place is with the scientific data on changes in Earth's temperature. It probably won't surprise you to learn that researchers are getting better and better at explaining the reasons behind shifts in Earth's temperature (figure 5.2). For example, atmospheric scientists attribute the brief drop in Earth's average temperature from 59.8 to 59.4 F in 1991 to the eruption of Mount Pinatubo in the Philippines. When this volcano blew, it ejected a fine mist of smoke, ash, and sulfate aerosols into the atmosphere. These ejecta mixed with the upper atmosphere and were then dispersed around the globe. Due to their relatively large size, these materials actually blocked out, to a degree, the sun's heat, thus causing the observed cooling. As these volcanic ejecta gradually fell out of the atmosphere, Earth's temperature began to increase again, starting in late 1992.

Figure 5.2 reveals that Earth's temperature also underwent a slight cooling and then stabilization from the 1940s to the 1960s.[8] This cooling was

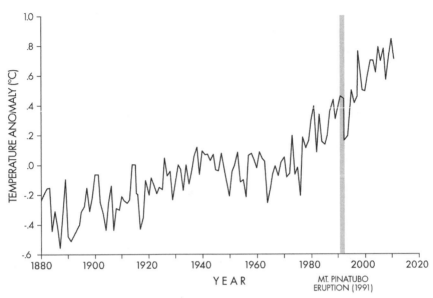

Figure 5.2. Earth surface temperature anomaly from 1880 to 2010. "Anomaly" here refers to departure from the long-term "stable" temperature value. A positive anomaly means that the observed temperature is warmer than the long-term reference value.

concentrated in the Northern Hemisphere and was likely the result of rapid industrialization there. Like a sequence of volcanic eruptions, this industrialization in the north triggered a sharp increase in haze (comprised of sulfate aerosol particles) that blocked the sun's warming radiation, leading to cooling. With the advent of the 1970s, and the enactment of air pollution abatement laws, atmospheric particulate pollution began to decline but carbon dioxide emissions from the burning of fossil fuels have continued to grow, decade by decade, and so has global warming. Indeed, according to NASA scientists, the first decade of the twenty-first century was the warmest in recorded history.

Though there is variation—with occasional breaks in warming—it appears that the overall trend toward a hotter planet has now been set in motion. This is the consensus of the prestigious Intergovernmental Panel on Climate Change (IPCC) established by the UN in 1988. The IPCC is charged with integrating research findings from more than two thousand climate scientists spread across more than 120 nations. The most recent report from these scientists confirms what some researchers had already begun to suspect back in the 1980s—namely: "Warming of the climate system is unequivocal"—meaning certain.[9]

DOES THE WARM DAY IN JANUARY
PROVE THAT GLOBAL WARMING IS REAL?

Weather changes hour by hour, day to day, and in the course of any given year there will always be days that are unseasonably warm, or cold. This means that a spell of warm weather in winter doesn't prove global warming is occurring any more than a period of unseasonably cool weather in summer disproves the reality of global warming. Weather, after all, is a short-term phenomenon (think hours, days, months); climate, on the other hand, is a long-term phenomenon (think decades, centuries, millennia). What is new today isn't weather change—that's a given; what's new is climate change!

Understanding the myriad factors that affect climate change is enormously challenging, presently engaging many thousands of scientists around the world. At Penn State alone, more than forty of my colleagues are involved in climate-change research. They occupy a variety of disciplines including geography, climatology, earth-system science, mathematics, biology, computer modeling, chemistry, economics, and physics. No matter their specific discipline, these scientists share a passion for discovery coupled with a deep commitment to learning the truth.

What Does Climate Change Look Like on the Ground?

You might think that global warming just means warmer days and more time at the beach. Think again: In its most fundamental expression, global warming means that the amount of energy in Earth's climate system is steadily increasing. By analogy it's like turning up the stereo volume, bit by bit, until you reach a point beyond what the speakers can handle. If you do this, the result is distortion and destabilization; and it's the same when we ramp up the energy in Earth's climate system. This is why scientists tell us to prepare for more climate change, not more sunbathing at the beach.[10]

Global warming is no longer a specter on some distant horizon, but a reality that is evident, not just to scientists, but to all people, no matter where we live, from the poles to the tropics.

Consider the Arctic

Scientists have known for some time that the Arctic was warming nearly two times as fast as the rest of the planet. Nonetheless, the announcement in the summer of 2007 that large zones of Arctic ice had melted away came as shock to many climate scientists. In fact, the area of Arctic summer ice cover has declined by 40 percent in recent years. The summer melt area now comprises some 1.3 million square miles (equal to more than one-third of the continental United States). This matters because previously 80 percent of the solar energy that hit the white surface of the frozen Arctic Ocean in summer was reflected back into space, but now, with annual massive summer ice melts, most of that heat is absorbed by the dark surface of the exposed ocean.[11] The result is both warmer water and a greater transfer of seawater to the atmosphere via evaporation, both of which are further shaking up Earth's climate.[12]

Consider the Tundra

Travel south from the Arctic Circle and you will enter into vast swaths of treeless tundra vegetated by lichens and mosses and stunted shrubs. This is the land of moors and bogs and mires. Below the thin surface soil layer is a zone of permanently frozen ground—sometimes hundreds or even thousands of meters deep—composed of carbon-rich organic matter that has been deposited over tens of millions of years. This material, referred to as "peat," is in the early stages of coal formation. Until recently, this organic matter was in a kind of *cold-storage locker*, but now, as the Arctic biome warms, the surface of this permafrost layer is beginning to melt in some places. When this occurs, water evaporation increases, leading to drying, and this in turn creates conditions favorable for microbes to break down the previously frozen organic matter, releasing carbon dioxide to the atmosphere in the process. Drying of the tundra also creates conditions for wildfires such as the one that incinerated 400 square miles of Alaska's arctic tundra in 2007. Conditions for such fires have not existed for the past ten

---

A WARMER WORLD IS A WETTER WORLD

Since the 1980s global rainfall has been increasing by about 1.5 percent each decade. This is simple physics; the warmer the air, the more moisture it can hold.[13] But as we would expect with the destabilizing effects of more energy in Earth's climate system, this does not lead to a uniform increase in rainfall everywhere. Some places, like North Africa, are now experiencing more frequent and severe droughts. Other places, like Southeast Asia, are being confronted with more frequent and severe storms and floods.

thousand years (i.e., since the end of the last ice age).[14] Whether it is by decomposition or fire, the result is the same: A large carbon pool that was once safely cloistered away in frozen darkness is now beginning to move into the atmosphere as CO_2 gas.

The conversion of peat to atmospheric CO_2 is also perched to take off in a big way in the sub-Arctic region of Western Siberia. Here, an expanse of once-frozen bog land (the size of France and Germany combined) contains an estimated seventy billion tons of methane, a greenhouse gas with twenty times the heat-trapping capacity of CO_2. This methane has heretofore been locked away within the permafrost, but now, with the onset of melting, it is bubbling up into the atmosphere.[15]

Consider the Forests of the American West

Forests play an important role in steadying global climate because, as they grow, they take carbon dioxide out of the atmosphere and store it in their wood. For example, the forests of the Western United States have been responsible, until recently, for about one-third of U.S. total carbon storage. However, Western forests—instead of serving as a *sink* for atmospheric carbon—are now beginning to act as a net *source*, intensifying, rather than lessening, global warming.[16] Indeed, up and down the Rocky Mountain chain, trees have been dying in unprecedented numbers. In Wyoming and Colorado there are now millions of acres of dead pine trees with more dying each year. The same is true farther north in British Columbia where tens of millions of acres of pine forests have been devastated. The culprit in this massive dieback is the pine beetle (*Dendroctonus ponderosae*). This beetle drills its way through the bark of western pines, taking up residence in the tree's sapwood. In the process, it introduces a fungal pathogen that further weakens the pines.[17]

This beetle has had a low-level impact on western pines for a long time, but nothing compared to what's happening now. So, what's changed? Certainly, climate for one thing. Prior to global warming, most adult pine beetles died with the onset of winter. That was good for the pines because the beetles had to build up their populations from scratch each spring. But now, under warmer conditions, adult beetles are increasingly able to survive through the winter and their populations explode during the summer with devastating effects on the pines. Add to this the fact that the pines, already stressed by a hotter and drier climate, are less able to defend themselves from the beetle onslaught.

The story doesn't stop here. All the dead trees, both standing and collapsed, represent an enormous stock of dry fuel. Add a spark and you can easily get a forest fire. And it turns out that global warming is contributing those sparks because as the planet heats up the frequency of lightning strikes—think match strikes—increases. On the ground this can translate to hundreds, even thousands, of lightning-initiated wildfires in a single day. Indeed, fully half of the U.S. Forest Service budget is now spent combating fires, leading some to dub it the U.S. Fire Service.[18]

Consider the Amazon Rain Forest

In recent times many people feared that chopping down trees would be the end of the Amazon rain forest, but now it appears that the threat of tropical deforestation may be small potatoes compared to the impacts of climate change—specifically drought and fire. Already the great Amazon forest is dying at its margins—morphing into savanna composed of drought-tolerant shrubs and wiry grasses. Some climate experts now predict that as much as 85 percent of this vast tropical jungle, extending across the belly of South America, could be reduced to open savanna if spiraling greenhouse gas emissions are not quickly brought under control. Even a modest rise in temperature above preindustrial levels (which now appears likely) could alter rainfall, drought, and fire regimes enough to eliminate one-third of Amazonia's forests.[19]

Finally, Consider Ocean Coral Reefs

Of all marine habitats coral reefs are the most enchanting and diverse, filled with a wide array of colorful fish species. The reefs themselves result from a plant-animal symbiosis. The animals (known generically as "polyps") construct the skeleton-like body of a coral reef through the gradual deposition of calcium carbonate. The plants (various types of algae) embed themselves in the coral body, receiving both shelter and the carbon dioxide they need for photosynthesis from the polyps. The polyps, in turn, benefit by receiving up to 90 percent of their food from the algae.

Though coral reefs have existed for millions of years, an estimated 20 percent have been lost since the 1980s and 60 percent may disappear by 2050.[20] Why? Again, climate change appears to be a big part of the reason—specifically, rising ocean temperatures plus ocean acidification. Both are related to human-induced rises in CO_2 emissions.

It only takes a small rise in ocean temperature—similar to what is now occurring—to weaken the plant-animal symbiosis of a living coral reef; and warming has already led to reef demise in many places.

Add to this the fact that fully one-third of human-generated greenhouse gas emissions is absorbed by the sea. Once in the ocean, this surplus carbon dioxide is converted to carbonic acid, increasing seawater acidity. Indeed, since the start of the Industrial Revolution—that is, when humans began burning fossil fuels in earnest—the world's oceans have absorbed an estimated five hundred billion tons of carbon dioxide, thereby raising average ocean acidity by 30 percent. This matters because as seawater becomes more acidic, coral reefs cease their growth and, unless there is a reversal in acidity, gradually deteriorate to rubble.[21] As famed oceanographer Sylvia Earle warns: "Alter the chemistry of the ocean [as global warming is now doing] and the entire system shifts. Some natural changes we can predict but it is impossible to anticipate how fast, or how much will occur as a consequence of tipping the ocean's chemistry onto a different course."[22]

Putting the Pieces Together

From the foregoing I hope it is clear that climate scientists are not simply cloistered in laboratories working with computer models. Climate modeling is built on real-world data—the result of millions upon millions of hours of laborious fieldwork, including temperature measurements in the oceans, forests, plains, mountains, air, and soil across every continent and in space, not to mention painstaking efforts to reconstruct past climates using tree rings and ice cores, as well as detailed monitoring of the effects of climate change on Earth's plants and animals.

For the sake of simplicity I described the effects of global warming in selected regions—Arctic, tundra, Western U.S. pine forests, Amazonia, ocean reefs—but, as we know, Earth operates as one interconnected system. This means that what happens in one place can, and often does, ramify throughout the entire biosphere, creating reverberations far and wide. For example, as touched on above, the disappearance of summer ice from the North Pole is heating up the Arctic's water and air in summer; and this added warmth is catalyzing a thawing of the permafrost thereby creating conditions for the decomposition of the carbon-rich organic matter that has been, up to now, locked away. With this new organic matter decomposition comes more CO_2 release and more warming. And this means warmer winters in places like the U.S. Rockies, enabling the populations of pests, like the western pine beetle, to explode and decimate forests, thereby creating conditions for massive forest fires and yet more carbon dioxide release to the atmosphere and still more warming and drying. And so it goes.

In sum, what may appear as isolated events—for example, the melting of Arctic sea ice—can, through reinforcing feedback mechanisms, make global warming a self-reinforcing process. In other words, global warming once it starts, can, and invariably does, lead to more of the same. The operation of these feedback mechanisms also means that climate change is not necessarily a slow and linear process. In fact, dramatic changes can occur abruptly and with startling speed. Also, it bears noting that after a certain point, even if humans do wake up to the danger, the various self-reinforcing feedback loops could override all our efforts to reverse the warming process.

In sum, climate scientists concur that Earth's climate is becoming more chaotic and that climate tumult will continue to intensify into the future. And this matters tremendously because our entire civilization is based on the expectation of relative climate constancy. Yes, we can tolerate an occasional drought or hurricane, but our lives will begin to unravel if Earth's climate continues to shift in the direction of frequent and unprecedented floods, horrendous hurricanes, catastrophic droughts, epidemic disease outbreaks, torrid temperatures, and accelerating sea-level rise. So, "global warming is no longer a philosophical threat, no longer a future threat, no longer a threat at all. It's our reality."[23] Yes, you and I, whether we like it or not, are in the midst of a giant planetary experiment.

FOUNDATION 5.2: JEOPARDIZING HUMANITY'S NATURAL CAPITAL

When most people hear the word "capital," they think of money, cash flow, investments, and so forth. But there is another form of capital that is far more fundamental than money—namely, natural capital. Natural capital refers to Earth's stocks of such things as fertile soils, ocean fishes, productive forests, fresh air, and clean water. In other words, natural capital calls our attention to the things that serve as the foundation for our quality of life. When natural capital stocks are depleted or compromised, as is now occurring, human well-being is jeopardized.

Jeopardizing Earth's Soil and Water Capital

Without fertile soil humankind could not survive. Farmers know that soil capital increases when organic residues like manure and compost are added

IT'S NOT JUST SOIL CAPITAL
THAT IS BEING SQUANDERED

Just as soil capital is being lost from U.S. farms, the same is true for water capital. In the United States the depletion of water capital is evident most dramatically in the case of the Ogallala aquifer, an immense underground water reserve located beneath the Great Plains states. The water in the Ogallala was deposited millions of years ago and remained there undisturbed until the advent of irrigation-based, industrial agriculture. Currently, more than thirteen million acres of farmland in America's Great Plains are irrigated with water from the Ogallala. New water enters this aquifer each year in the form of rainfall, which slowly percolates down through the soil; but irrigation water is being mined from the Ogallala much more rapidly than new water is being deposited. In fact, since irrigation began in the Great Plains in the 1940s, water levels in the aquifer have declined by more than one hundred feet; and at the current rate of extraction (2.7 feet/year), this aquifer will be essentially drained within the next several decades.[24]

to the soil; soil capital also increases, albeit slowly, as rocks in the soil undergo physical and chemical weathering, releasing minerals to the soil. Just as soil capital can accumulate so, too, it can be depleted, as when topsoil is eroded away by wind and/or water. Ideally, new additions of soil capital will either balance or exceed soil losses. However, if more soil is eroded away each year than is formed, soil stocks will decline. Think of it as money in your bank account. If you are always taking money out and seldom putting new money in, you will eventually go broke. It's the same with soil!

The United States Department of Agriculture has been monitoring the status of U.S. soil stocks for many years. Their data reveal that, on average, four tons of soil are lost per acre each year from U.S. croplands. In fact, in the United States, soil is now eroding ten times faster, on average, than it is forming through natural processes. This means that, on average, one inch of topsoil is lost from U.S. cropland every thirty years.[25] It is important to underscore the word "average." On some farms, the loss of soil is greater than this average; on others, the loss is less; and on a small number, there is actually a net gain of soil because of conscientious soil-building and conservation practices.

Of course, if soil capital stocks were huge, it would not much matter if these stocks were overdrawn a bit each year. Unfortunately, though, U.S. soil stocks are not huge. For example, already the farm state of Iowa has lost more than half of its fertile topsoil, and as much as 75 percent of the soil has been lost from some Midwest farms.[26]

In many parts of the world, average soil erosion rates are even higher than in the United States. For example in China and India (comprising roughly a third of the world's population), soil is eroding 30–40 times faster than it is forming; and worldwide more than twenty-five million acres of cropland are permanently removed from farming each year due to the cumulative effects of soil erosion.[27]

Jeopardizing Human Health

Just as fertile soil and abundant fresh water represent important natural capital stocks, a healthy, robust human being is also an important font of natural capital insofar as s/he can create and produce wonders. But, just like soil capital, human capital in America (when gauged by the metric of physical and emotional well-being) is now being compromised by poor diet, stress, sedentary lifestyles, depression, heart disease, obesity, cancer, diabetes, and respiratory ailments, not to mention a deteriorating natural environmental.

Man-Made Chemicals Causing Human Cancers?

The spread of man-made chemicals throughout Earth's land, sea, and air is an especially grave threat to human well-being. More than one hundred thousand different man-made chemicals are now in circulation on planet Earth, with an average of two to three new synthetic chemicals (with largely unknown health and environmental effects) released into the global environment every day.[28] The vast majority of these chemicals have been invented within the last hundred years, most since 1950. As a result, there are now billions of tons of synthetic chemicals free in the environment and every one of us alive today has more than two hundred different synthetic chemicals lodged in our body tissues.

This matters insofar as numerous man-made chemicals are now classified as possible cancer-causing agents, or carcinogens; and the incidence of many human cancers has been rising in recent years. In some cases, this increase is partially the result of improved technologies for the early detection of cancer. For instance, prostate cancer can be detected more effectively now than in the past. But in the case of childhood cancers, which have increased significantly since 1970, early detection technologies are not employed. Early detection is also not a factor in the rise in testicular cancer for young men between nineteen and forty years old (a threefold increase since 1940); nor is it a factor in the documented rise in brain cancers among both children and the elderly.[29]

The load of foreign chemicals that each of us carries in our body tissues—and, by extension, the probability that any one of us might suffer from an environment-related cancer—depends, in part, on where we have lived during our lifetime. For example, cancer-mapping studies often reveal a correlation between the abundance of certain categories of chemicals in the environment (e.g., pesticides, industrial solvents, radio-isotopes) and the incidence of certain types of cancers. So it is that breast, bladder, and colon cancers are most common in regions that are highly industrial (e.g., Eastern seaboard, Great Lakes Basin, Baton Rouge region); while non-Hodgkin's lymphoma is most concentrated in agricultural regions (e.g., Great Plains and up into Minnesota and Wisconsin).[30]

If you are curious about the chemical loads in the place where you live, just visit www.scorecard.org and type in your zip code and this site will tell you the chemicals that are in your county's air, water, and soil—and, by inference, possibly in your body.

When I visited Scorecard, I learned that my home county (Centre County, Pennsylvania) is among the dirtiest 30 percent of U.S. counties as regards carbon monoxide, nitrogen oxide, and sulfur dioxide releases and overall

---

A DILEMMA FOR NURSING MOTHERS

The proliferation of foreign chemicals in the environment is taking some of the joy out of motherhood. For example, take the case of a young woman who has just given birth to her first child. She knows about the advantages of breastfeeding—how it will fortify her newborn's immune system, helping to ensure that her infant will have fewer bouts of sickness. But what she hasn't heard very much about is the chemical contamination of her breast milk. Indeed, of all human food, breast milk is now among the most contaminated. Consider: When this young mother nurses, fat globules throughout her body are mobilized and carried to her breasts, where they are transformed into milk. Toxins that have slowly accumulated in the mother's body fat over the years will be part of those mobilized fat globules, transformed to milk and passed on to her infant.

The young mother faces an obscene dilemma. What is she to do? Forgo the many advantages of breast milk for her child by feeding her inferior infant formula? Or fortify her child's immune system with her breast milk while simultaneously contaminating her baby with the synthetic chemicals stored in her own body fat? This is not a decision any woman should be asked to make.[31]

cancer risk from hazardous air pollutants. I also learned that local industries annually release more than twenty tons of zinc compounds into my county's environment, and that zinc is a suspected developmental, immunological, reproductive, and respiratory toxicant. What about where you live?

The rise in human cancers has been paralleled by a rise in cancers among animals, especially since 1960. These animal cancers—for example, clams with gonadal cancer, fish with tumors—are almost always associated with chemically contaminated environments, such as polluted bays. Animal cancer studies are highly relevant to humans because many of the genes that govern cell division (i.e., those associated with cancer) are conservative—that is, they are the same (or very similar) in humans as in other animals. Moreover, animal studies are easier to interpret than those involving humans because animals don't smoke, eat fatty food, experience work-related stress, and so forth. In other words, animals don't introduce the confounding variables that make it so difficult to unequivocally pin down causality in human-cancer studies.[32]

Man-Made Chemicals Derailing Human Development?

It wasn't until the 1980s that a significant number of researchers began to entertain the possibility that man-made chemicals—in addition to sometimes being carcinogens—could also interfere with human embryogenesis—the process by which a fertilized egg in a mother's womb develops into a fully formed baby. But it is now known that a surprising number of synthetic chemicals can inadvertently scramble the hormone signals necessary for the correct development of the human fetus.

Normally, as the human fetus develops, hormones direct critical events, such as sexual differentiation and the proper construction of the brain. Correct development depends on getting the right amount of the right hormone to the right place at the right time. If the hormone messages don't arrive or if they arrive in the wrong amounts, the embryo's development can be derailed (figure 5.3).[33]

Scientists have already identified more than one hundred individual chemicals or chemical families—all man-made—that disrupt natural hormone expression in humans. These chemicals act by either mimicking the body's natural hormones or by blocking the action of the body's natural hormones. Such chemicals have been dubbed "endocrine disruptors" because they interfere with the normal functioning of specialized endocrine system organs—such as the pituitary, thyroid, ovary, and testes—that produce, store, and secrete hormones.

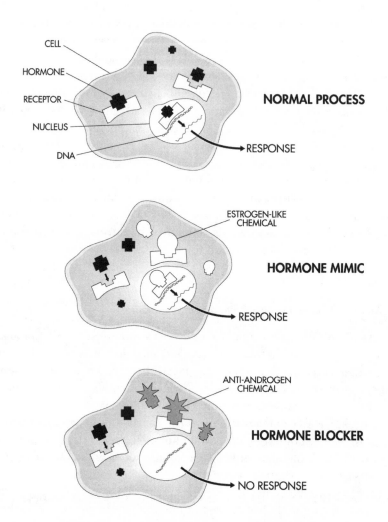

Figure 5.3. Hormones trigger cellular activity by hooking up—lock-in-key fashion—to hormone receptor sites, thereby activating genes in the nucleus (top cell). Foreign chemicals that mimic hormones, should they be present in the developing embryo, can bind with hormone receptor sites (e.g., middle cell) or block these sites (hormone blocker—bottom cell), thereby disrupting the normal unfolding of embryo development.

ENDOCRINE DISRUPTORS AND WILDLIFE

Numerous studies of amphibians, birds, and mammals have documented many types of severe developmental abnormalities, such as extra limbs in salamanders, shrunken penises in alligators, and eyes developing in mouth cavities in frogs.[34] Add to these the recent discovery that some male fish—for example, small-mouth bass—in many U.S. rivers are now producing eggs. That's right: Male fish producing eggs!

Evidence is mounting that endocrine disruptors—specifically estrogen-mimics—may be at the root of this so-called gender bending. Fish biologists report that 80 percent of the streams tested in thirty states contain synthetic estrogen and its mimics. How has this happened? One avenue is runoff from farms. Estrogen-mimicking chemicals are often present in the tens of millions of pounds of pesticides sprayed on U.S. farms each year. Only a tiny fraction of this spray actually lands on pests; the excess ends up in farm soils and nearby waterways. Add to this the estrogen-like compounds now contained in many soaps and cosmetics and the synthetic estrogen in birth-control pills and estrogen supplements—the great majority of which eventually gets washed down drains and toilets, making their way to sewage treatment plants and usually from there into surrounding waterways. Perhaps it is no coincidence, then, that feminized-male fish are especially common in stream stretches immediately below wastewater disposal facilities.[35]

When our great-grandparents were born endocrine disruptors were virtually nonexistent. Now, they are everywhere—e.g., in detergents, toys, food-can liners, toothpaste, garden hoses, laptops—and, not surprisingly, in our very bodies. And this fails to account for the tens of thousands of synthetic chemicals, now free in the environment, yet to be screened for their endocrine-disrupting potential.

Grasping the Ramifications of Endocrine Disruption

Recently, when a team of scientists reviewed 800 scientific studies on the effects of endocrine disruptors they concluded that it is "remarkably common" for very small amounts of hormone-disrupting chemicals to have profoundly adverse effects on human health.[36] This is not so surprising insofar as the endocrine system is the master control center of the body, determining, to a significant degree, the proper functioning of living organisms.

It is still too early to grasp the full significance of this endocrine disruptor saga. Some scientists believe we are still only seeing the tip of the proverbial iceberg. For example, it may turn out that things like the sharp uptick in human obesity and diabetes in recent decades is tied to endocrine disruptors. Indeed, when mice in utero are exposed to low doses of endocrine disruptors, they are born with normal levels of fat, but as they mature to adulthood they develop excessive body fat.[37] It also might be the case that hormone disruptors are playing a part in the early onset of puberty, especially in girls. For example, breast development is now starting at seven to ten years of age versus thirteen to sixteen years of age before the widespread introduction of hormone-disrupting chemicals into the environment. Of course, there could be other factors at play, but clearly something is amiss. Of all the possible worrisome consequences of endocrine-disrupting chemicals, reports on reductions in the human male sperm count (i.e., the number of spermatozoa per milliliter of semen) have garnered special attention (and concern). Biologist Michael Zimmerman explains why:

> A frightening collection of recent studies has begun to suggest that emasculation is taking place. . . . A massive study published . . . by the *British Medical Journal* . . . found that average sperm counts have decreased 42% from 113 million to 66 million sperm cells per milliliter, within the past fifty years.[38]

There is a growing research effort aimed at determining what exactly is going on here, and it may well be that events during embryogenesis are at the root of this decline in the male sperm count. Here is one scenario that could explain what is happening. First, it is well known that from the moment of conception until the human fetus is almost two months old it is impossible to tell if it will be a male or a female. And then, on a certain day, small amounts of a certain hormone kick in, instructing receptor sites—located on an embryo tissue patch known as the "gonadal ridge"—to begin to develop into either testicles or ovaries. But this normal unfolding of gender could be confused if foreign chemicals that mimic or block natural hormone messages are present in the fetus. For example, let's say that the hormone testosterone that specifies male genitalia arrives to receptor sites on the gonadal ridge. Once present, this testosterone sends the signal to create special cells called "sertoli cells" that later on (when the male child reaches puberty) will become the site of sperm production. Given the fact that adult males with lots of sertoli cells are able to produce lots of sperm while those with fewer sertoli cells produce less sperm, biologists have speculated that for males with a low sperm count, something may have happened early on, when they were developing in their mother's womb, that limited sertoli cell formation. For example, perhaps a large portion of the testosterone receptor sites on the gonadal ridge were blocked by a foreign hormone-like chemical—and this resulted, ultimately, in an adult male with fewer sertoli cells in his testes and correspondingly lower sperm production potential. This stands as a working hypothesis; the jury is still out.

STEPPING BACK TO SEE THE BIG PICTURE

> In order for us to maintain our way of living, we must, in a broad sense, tell lies to each other, and especially to ourselves. . . . The lies act as barriers to truth. These barriers to truth are necessary because without them many deplorable acts would become impossibilities.
>
> —Derrick Jensen[39]

The unprecedented production and release of synthetic chemicals into the environment—like the unprecedented release of greenhouse gases—constitutes an unpremeditated planetary experiment that we humans have set in motion. Here is the experiment: For tens of thousands of years humans lived in an environment of naturally occurring chemicals; then, about a hundred years ago, we began synthesizing tens of thousands of chemical formulations—completely new to our evolutionary experience—and spreading these chemicals, by the millions of tons, throughout Earth's air, water, and soil. Given the magnitude of this experiment, perhaps we should not be surprised that we are now experiencing a proliferation of previously rare cancers, that the male sperm count is apparently declining, and that developmental abnormalities are on the rise.

Our recklessness has implications for the entire community of life. For example, consider the lowly honeybee. Many regard bees as annoyances, but what if I told you that, on average, every third bite of food you put into your mouth comes to you by the grace of bees? That's right: Honeybees play a critical role as pollinators for about one-third of all agricultural crops. Think: apples, pears, peaches, almonds, cucumbers, broccoli, onions, pumpkins, carrots, avocados, blueberries, cantaloupe, asparagus, celery, watermelons, soybeans, sunflowers, grapes, olives, raspberries, strawberries, citrus. . . . You get the idea. No bees, no pollination; no pollination, no fruits, no seeds, no nuts.

And it's not just our food plants that need pollinators. Most wild plant species require insect pollinators to ensure reproduction. Moreover, without pollination, most wildlife would no longer have access to the flowers, seeds, and fruits they depend on for their survival. It is no exaggeration to say that pollinators are critical to life on Earth. If life is relationship, it is pollinators that are the great connectors. And this brings us back to honey-bees and their well-being.

Starting in 2006 some U.S beekeepers began reporting that their hives were dying for no clear reason. One week the bees would be fine and the next they would literally disappear from the hive, never to return. There would be honey in the hive and larvae, but the adults would be gone. This strange phenomenon, now referred to as "colony collapse disorder," has also been occurring in Europe and Asia.

Given the importance of honeybees to both human crops and the plant kingdom in general, researchers have been working to solve this colony col-lapse mystery. A variety of biotic factors, including parasitic mites, viruses, fungi, and bacteria, have been considered but none, by itself, appears to be the culprit. That said, there is growing evidence that a particular class of agricultural chemicals—neonicotinoids—that are now widely used to kill insect crop pests may be the primary catalyst for bee colony collapse. Neonicotinoids are nerve toxins that went into wide use on U.S. farms just a few years before the first reports of colony collapse disorder. Prior to their approved release by the USDA, these chemicals were not tested for their possible deleterious effects on non-target insects like honeybees.

In practice, neonicotinoids are applied directly to crop seeds as well as to farm soils and become incorporated into both crop and wild-plant tissue, including the pollen and nectar that bees forage for and take back to their hives.

Based on a battery of studies, it is now known that neonicotinoids weaken the immune system of bees, making them more susceptible to fungal infec-tions and parasite attacks. Neonicotinoids have also been shown to interfere with bee learning and navigation. This is not a trivial matter for creatures that travel long distances from their hives on their daily foraging forays. Indeed, French scientists have shown that honeybees subjected to neonic-otinoids have a much higher probability of getting lost—that is, not finding their way back to the hive—compared to bees not subjected to these chemi-cals. Furthermore, a recent study by researchers at the Harvard School of Public Health found that hives given high-fructose corn syrup (a common practice of commercial beekeepers) had a much higher incidence of colony collapse when the corn syrup was tainted with traces of imidacloprid (one of the most widely used neonicotinoids) compared to corn syrup free of any trace of imidacloprid.

Of course, honeybees aren't the only pollinators out there. Butterflies pollinate and so do moths, not to mention beetles and flies and bats, but honeybees are, by far, the most important when it comes to human food crops. We know the most about them and what we are learning may fore-shadow similar consequences for other insect pollinators.

In sum, it is not just human reproduction that may be in jeopardy as a result of the widespread broadcasting of new chemical formulations; sexual reproduction in the plant world—upon which the entire community of life depends—could also be at risk. Indeed, though they appear different from us, the bees' fate and our fate are intertwined.[40]

WRAP-UP: EXPANDING ECOLOGICAL CONSCIOUSNESS

> It is the destruction of the world in our own lives that drives us half insane, and more than half. To destroy that which we were given in trust: how will we bear it?
>
> —Wendell Berry[41]

This chapter underscores one of this book's central messages—namely: When it comes to planet Earth, it is impossible to do just one thing. Yes, everything we do, it seems, has repercussions. There is a tendency to minimize these repercussions by referring to unforeseen consequences as "side effects." For example, if you buy a container of pesticides you will likely see an advisement on the back warning you of possible side effects, but in truth, there are no side effects, only effects!

The effects of human actions can become especially pernicious when we add too much of something dangerous—like carbon dioxide or hormone-disrupting chemicals—into the body of Earth. By the same token, we also get into trouble when we remove too much of something essential—like fertile topsoil or aquifer water—from its proper place in Earth's body.

It is becoming increasingly clear that if we continue to barrel ahead—taking without giving back—we will leave a tattered planet for future generations. It is even conceivable that there may be *no* future generations; humankind may not make it through this difficult period. Just as other species have gone extinct, we humans may now be moving toward our own extinction. It is important that we, at least, acknowledge this possibility.

Of course, we still, hopefully, will have some say in this matter. We can help to avoid the grim specter of our own extinction by, among other things, adopting the Precautionary Principle which states that "an action should not be taken if the consequences are uncertain and potentially danger-ous."[42] This principle reminds us that—in situations where there is both scientific uncertainty and a reasonable suspicion of harm—we should stop what we are doing until we are sure it is safe. Certainly, as regards rising atmospheric carbon dioxide levels and the proliferation of toxic chemicals in the environment, one need muster only a modicum of ecological awareness (and common sense) to conclude that "consequences are uncertain and potentially dangerous."

APPLICATIONS AND PRACTICES: COURAGE

> In our modern culture men and women are able to interact with one another in many ways: They can sing, dance or play together with little difficulty but their ability to talk together about subjects that matter deeply to them seems invariably to lead to dispute, division and often to violence.
>
> —David Bohm, Donald Factor, and Peter Garrett[43]

We live in a mythic time. Never before, it seems, have humans wielded so much power for both good and harm. We can annihilate each other in war and poison Earth, but we also have the potential to embrace one another in peace and to restore Earth to wholeness. Do you believe this? If you don't think this is possible, you are not unusual. Chances are good the person next to you doesn't believe it either. Indeed, many of us, around the world, have been conditioned, to varying degrees, to be spectators (allowing the media to interpret the world for us), rather than participating directly in the creation of a world of peace, sanity, beauty, and joy. We are told that we live in a democracy, as if simply being able to vote for simulacra candidates constitutes a democracy. Our collective disenchantment and disempowerment are evidenced by the fact that, in the United States at least, many citizens don't even bother to vote. But, believe it or not, common, everyday citizens have enormous latent power. The first step to uncapping this power is to muster the verve and courage to speak what we know to be true, first to ourselves and then to each other.

The Power of Truthspeaking

In the previous chapter I stressed that it is not enough to simply learn the facts about the beleaguered condition of Earth; it is also important to listen and, in so doing, to feel those facts—to take them in, somatically. This, I believe, is a necessary prerequisite to being able to access and then speak our deep truth in the face of Earth's decline.

Teacher and writer Tamarack Song calls on all of us to practice "truth-speaking," which, in his words, means " to state clearly and simply what one thinks and feels. There is no judgment or expectation, no disguise of humor or force of anger. The manner of speech is sacred, because it wells up from the soul of our being rather than from our self-absorbed ego."[44] When we truthspeak, we let go of our fears about how the other might interpret or react to our words. We simply speak our thoughts and feelings with utter honesty, absent any filtering, and, in so doing, experience genuine freedom. Note: Just the fact that you might pass a day without saying anything to another that would be judged as an outright lie does not mean that you have practiced truthspeaking, because truthspeaking, at its core, is about not betraying—not lying to—ourselves.

Thus, a key requirement for the practice of truthspeaking is that one be able to make distinctions between pseudo-truths—for example, knee-jerk opinions—and genuine truths. To illustrate this difference, I will share a story with you about Jason, a Penn State student who was taking a course on Buddhism. As part of the course Jason had an assignment to speak nothing but the truth for forty-eight hours. When I talked to Jason during the second day of this assignment he said that for the first time in his life he was really speaking the truth. Curious, I asked if he could give me an example of a *truth* he had spoken that day.

He responded: "I was just with my roommate and he looked bad and so I told him the truth—I said: 'You look really bad, man.'"

Hearing this, I had the uneasy feeling I get when I sense that something is not quite right—that truth is not being served. After pausing to collect my thoughts, I asked Jason: "So what do you suppose was behind your statement to your roommate that he looked bad? In other words, what was your deeper, personal truth in that moment?"

Jason looked perplexed, so I asked my question differently: "What did your judgment of your roommate—that he looked bad—reveal about you? What do you suppose you were needing in that moment?"

This time Jason responded, albeit sheepishly: "I guess I was needing him to not look bad?"

"And this means that . . . ?" I followed.

"It means that appearances are important to me," responded Jason.

For me, this question: "What does my response/reaction/judgment say about me?" is a necessary prelude to truthspeaking, for to speak the truth requires that we speak our own inner truth—that we be radically honest with ourselves—and this takes courage.

Song contends that all of us know how to truthspeak. We simply have forgotten. Indeed, we get derailed from truthspeaking anytime we get tangled up in shame and guilt and pleasing others and judgment and competition and control and fantasy and regret—which is to say, much of the time!

Insofar as truthspeaking takes enormous courage, it is not surprising that we use all sorts of techniques to avoid it. For example, sometimes we escape by using humor to sidestep touchy topics. On other occasions, we might use violent language to cover up the truth. For example, we may lash out with "F--k you" instead of saying what is true—such as, "I don't agree with you." Or we might say "Go to hell," instead of what is true—for example, "I won't do it."[45]

Perhaps the biggest obstacles to truthspeaking are small talk and gossip. Song advises:

> If you don't have anything to say, don't say anything. . . . Every breath is Sacred; every breath we are given is for a purpose. . . . If every breath is for a reason, then every word it carries is for a reason. Idle chitchat is like a waste of sacred breath. Using it to spread gossip or manipulate is like pouring vinegar in a trusting baby's mouth. . . . . When we resort to small talk, we are not speaking the Truth of the Now. We are merely filling this silence with fluff and drivel. We are trashing the Moment.[46]

Truthspeaking, in our time, is particularly difficult to the extent that we have been conditioned—in our families and schools and through the media—to dilute our personal truth to ensure that we will be accepted by others, imagining that if we told the truth, moment by moment, we would be shunned. After decades of suppressing our personal truth we may come to a point where we actually lose touch with what is true for us. We, in effect, go dead, becoming a patterned behavior—plastic people—rather than spontaneous, creative beings. If our speaking is dead, detached, manipulative, inauthentic, then this is, quite literally, the energy we broadcast out into the world, for everything we do has an effect—our every action is important.

And what does truthspeaking have to do with an environmental issue such as climate change, you might ask? A lot, I think, especially given all the deliberate obfuscation that misleads many into believing that climate change has no scientific basis. Truthspeaking is also in order when it comes to toxins in the environment. We can no longer afford to delude ourselves, imagining that we are somehow immune to the pernicious effects of our chemical spewings. Our well-being and that of future generations may be at stake.

Overall, truthspeaking is essential for the creation of a world of sanity, integrity, and well-being. Like everything else in this book, it is in the doing that we learn and grow. So, here are three practices that I have used for the cultivation of truthspeaking.

i: Truthspeaking by Treating Each Word as Special

You have, no doubt, heard the expression "Less is more." This seems counterintuitive. How could less be more? Well, for starters, call to mind people you have encountered in your life who speak few words but offer great wisdom. Compare them to those who ramble and rattle on in seemingly endless prattle.

There is a practice, dubbed "If I Only Had Five Words," that illustrates the power of lean, heartfelt responses. The rules are straightforward: Sit facing another person and simply take turns asking and answering each other's questions. Avoid questions that are superficial and, instead, craft questions that invite a deep and soulful response. Now, here's the most interesting part: You only have five words to answer each other's questions. You need not respond in complete sentences. For example, if your partner asks, "What makes your heart ache about being alive in these times?" wait with curiosity and patience until five words emerge from the core of your being that are your truth.

Less really can be more! If you muster the courage to genuinely engage in this practice, you will experience directly what it means to truthspeak.

ii: Truthspeaking to Ourselves

In my experience, it is exceedingly rare to find a human being who doesn't pretend in some ways, at least some of the time. Oftentimes the pretending I do is subtle, meaning that I may not even be aware that I am pretending. How about you? For example, do you sometimes pretend that:

- You have it all together.
- Everything is OK when it really isn't.
- You are happy when you really aren't.
- You like certain people whom, in actuality, you find irritating.
- You know more than you really know.

One way to explore this human tendency to pretend is to get a blank sheet of paper and write down "I pretend that _____" ten times along the left margin; and then, in the interest of truthspeaking, complete each of your ten "I pretend _____" statements. Make no mistake: This is not easy; it requires you to have the courage to be honest with yourself.

After completing your list, look over your statements and pick one that is particularly troubling to you. For example, perhaps you pretend to like someone whom you don't really care for. Circle the statement—whatever it is. Then, consider the following four questions:

1. How does pretending in this way serve you? After all, you wouldn't be pretending unless you *believed* it served you (benefited you) in some way!
2. What is the personal cost you pay for holding your belief that has you pretending?
3. In what ways might your life change if you were to let go of this belief that has you pretending?
4. What is a new possibility—a new belief—that you could substitute for your old belief and, in so doing, move from pretending to authenticity?

Again, none of this is easy, but it is worth the effort. Being honest with ourselves is a gateway to freedom—especially freedom from self-deceit.

iii. Truthspeaking for a Day

When you are ready to dip your toes a bit deeper into the water of truthspeaking simply make a pledge to truthspeak just for one day. To ensure that you get off to a good start, begin with a vow of silence. Consider what this means. What will it be like to make this vow, to not speak, to use your mouth simply for breathing instead of words. But cut yourself some slack by adding this amendment to your vow: "I will allow myself to speak provided I am sure that I am speaking my truth."

Then, when the inclination to speak arises, simply ask yourself, "What is true for me right now?" Ask it and mean it. Ask it and then go inside for the answer: What is true in my head? What is true in my body? What is true in my heart? Though many of us have long forgotten how to speak from a place of unfiltered presence, we can recover this ability. Pause right now, if you dare, and ask yourself, "What is true for me in this very moment?" Really stop and ask the question: "What is true for me right now?" Can you do it? It is not easy. Be patient with yourself. The first step, I believe, is simply cultivating the desire to know the truth.[47]

In sum, the world we now live in suffers from a dearth of courage and truth, and it is for this reason that truthspeaking has such latent power. Allow yourself to imagine a world where each of us has the courage to speak what is really true for us in all realms of our lives. Imagine the conversations we would be having if we weren't addled by insecurity, fear, judgment, confusion, and guilt. Substantive change toward peace and healing becomes possible when people change the way they relate to themselves and each other. Think about it: If you want positive change in your family, the way forward is to change the conversation—your way of relating to your family. After all, when you get right down to it, a family is a conversation. It's the same for the neighborhoods where we live. If we hope to change our communities (and the world!) we can begin by changing the way we speak to one another. No more hidden agendas, no more manipulation, no more displacement behaviors, just the heartfelt truth.[48] All of this is to say that the future doesn't just happen; rather, it is created—day by day—in no small part, as a result of our courage to speak our deepest truth.

QUESTIONS FOR REFLECTION

- What eyewitness evidence can you offer from your own life regarding climate change?
- Imagine that because of catastrophic climate change, starting tomorrow, you have to cut your energy usage by half—i.e., you will have to use only half the electricity, half the gas, half the heat, half the living space, half the stuff—that you are now accustomed to using. In what ways would your life be diminished, as well as enhanced, if this were to come to pass?
- Based on your perusal of scorecard.org, what is the situation regarding chemical pollutants in the place where you live and how does this information sit with you?
- It is possible for humankind to survive without access to capital in the form of money, but without access to *natural capital* we are all doomed. Do you agree or disagree with this statement? Why?
- What does it feel like for you to be a kind of *guinea pig* in a largely unpremeditated planetary experiment involving possible chemical havoc and catastrophic climate change?
- In what ways do you avoid truthspeaking? What price do you pay for diluting your truth? What benefits might accrue if you fearlessly spoke your truth?

6

LIVING THE QUESTIONS

Discovering the Causes of Earth Breakdown

> Once you have learned how to ask questions—relevant and appropriate and substantial questions—you have learned how to learn and no one can keep you from learning whatever you want or need to know.
>
> —Neil Postman and Charles Weingartner[1]

I have spent four decades investigating the condition of planet Earth. In the process I've asked lots of questions. Sometimes, my questions have involved field research. For example, during the 1970s and 1980s, noting that large swaths of rain forest were being hacked down and burned in the Amazon basin, I wondered if a new forest—complete with native trees and vines and insects and birds and mammals—could somehow reestablish on the ashen remains of the old. It turned out that in some cases the answer was "yes," but in other cases recovery was much less certain. This led me to ask *Why?* and to dedicate two decades of my life to pursuing answers.

Questions have also served as a catalyst in the writing of this book. For example, this chapter is centered on the question, What are the underlying causes of the myriad ecological upsets described in the preceding two chapters—including deforestation; songbird declines; species extinction; ocean degradation; climate disruption; soil loss; aquifer depletion; and the widespread chemical contamination of Earth's air, water, and soil (and by extension, our very flesh and blood)? I have not been satisfied to simply learn *what* has been happening—I want to know why! This quest to understand the *why of things* is akin to peeling back the layers of an onion, with the possibility of deeper understanding emerging as each successive layer is removed.

The desire to understand "why" is, I believe, fundamental to human nature, as illustrated in the following story about Primo Levi, a Jewish man forced to endure a long journey in a cattle car destined for a concentration camp during World War II. Deep into the journey, the train lurched to a stop and Levi, who was very thirsty, spotted an icicle hanging from a bridge abutment. He reached out to break it off but before he could bring it to his lips a hulking guard grabbed his arm and snatched the icicle away. Dismayed, Levi looked at the guard and asked, *"Warum?"* ("Why?"). The guard responded, *"Hier ist kein warum"* ("There is no 'why' here").[2] This story suggests to me that banishing "Why?" from our lives might be tantamount to extinguishing the core of our humanity. After all, the search for meaning is fundamental to human existence and there is arguably no better lever to guide us toward truth than the question "Why?"

FOUNDATION 6.1: HUMANS—TOO MANY, TOO MUCH?

Presently, in the early decades of the twenty-first century, Earth's human population is growing by more than 8,000 people per hour, 200,000 people per day, and more than 1.4 million people per week. At this rate, the human population will soon exceed 8 billion.

So here we are, already almost eight billion strong and with enough momentum to reach nine or ten billion before the end of this century. So far this may read as just numbers, but consider this: It wasn't until around 1850 that the human population reached one billion people for the first time. Mind you, we'd been around as a species for more than a hundred thousand years and it took all of that time to reach a population size of one billion! But, then, in a matter of only eighty years the human population doubled to two billion. That's right: In just eighty years—from 1850 to 1930—we did what had taken the previous 100,000 years to achieve. And it didn't stop at two billion: In the forty-year span from 1930 to 1970 our population doubled again, from two to four billion; now our numbers are doubling a second time, to eight billion and beyond.

If you are following me, you may be wondering how it is even possible for a population to double in size so quickly. To understand how, imagine an island with a human population of 2,000 people and a population growth rate of 5 percent per year. If our hypothetical island has 2,000 people at the beginning of the year, its population will have grown to 2,100 people by the end of the year (0.05 × 2,000 people = 100 new people added to the population). After two years the population will have grown to 2,205 people (0.05 × 2,100 people = 105 new people). If you continue with these simple calculations, you will discover that in fourteen years the island's population will have grown to four thousand people. In other words, given a 5 percent population growth rate, the population will have doubled, from two to four thousand, in just fourteen years! And, given fourteen *more* years, this hypothetical population would double again, to 8,000 people. Indeed, as long as the growth rate stays at 5 percent, the population would continue to double every fourteen years.

Using the so-called Rule-70, it is possible to quickly calculate the doubling time of any population because the doubling time is simply the number 70 divided by the percent growth rate. Hence, in our example 70/5 = 14, the number of years it took the island population to double.

OK, we have addressed the *how* of population growth, but not the *why*—specifically: Why has the human population been exploding since the 1800s? If you are like most folks, you probably assume that the reason is simply that women are having more babies these days than in the past. But it's important to question assumptions—especially ones like this one that may seem self-evident. Based on extensive demographic research, it turns out that the recent surge in human population growth is primarily the result of a decline in the death rate, *not* an increase in the birth rate. Indeed, in the centuries prior to today's population surge, women were still birthing roughly the same number of babies, but starting in 1800s proportionately more of those newborns were surviving the challenges of birth and early childhood. The same was true for mothers—that is, childbirth became less risky. The primary explanation for this is modern medicine—especially the realization that sanitation measures (as simple as hand washing) could greatly reduce the spread of microbe-borne infectious diseases. A further significant increase in survival resulted from the development of vaccines, particularly those aimed at ridding the world of once deadly childhood diseases like measles, whooping cough, polio, and diphtheria. More babies surviving to adulthood meant more potential mothers and, in turn, more new babies.

As this book goes to press, the human population growth rate is hovering around 1 percent. So, how long might it take for our numbers to double to fifteen or sixteen billion? Rule-70 predicts seventy years (70/1 = 70). But this calculation is based on the assumption that the current population growth rate will stay at 1 percent over the next seventy years; and

WHY DOES FAMILY SIZE TEND TO BE
LARGE IN LESS-DEVELOPED COUNTRIES?

Among those living in "developed" countries (e.g., United States, Western Europe, Japan), there is a tendency to blame those in so-called less-developed countries (LDCs) for the human population explosion. While it is true that people in LDCs often have large families, seldom do we ask, "Why is this so?" So, take a moment, now, to imagine yourself as a member of a cash-poor family living, say, in the highlands of Ecuador. You are twenty, the eldest among five children. Your family has no electricity and you rely on wood to warm your modest shelter and to cook your food. Your days, and those of your siblings and parents, are spent engaged in such activities as creating and repairing terraces for your crops; cultivating, weeding, and harvesting vegetables and grains; processing and preparing food; herding goats and making cheese; weaving clothes and making handicrafts; constructing and repairing tools; gathering wood and water from distant sources; not to mention caring for one another. If you had the wherewithal to construct a time budget—an accounting of how the seven people in your family spend their time—you would realize that having lots of kids makes sense. It means more hands and legs and backs and heads to help with the many time-consuming tasks necessary to simply run a self-sufficient household.

in actuality the growth rate has been declining slowly in recent decades, from a high of 2 percent per year in the 1960s (Think: thirty-five-year doubling time!) to its present 1 percent. Most demographers now predict that the worldwide population growth rate will continue to decline with human numbers topping out at between eight and nine billion toward the end of this century. We shall see. In the meantime, it is heartening to know that there are already some countries—such as Japan, Italy, and Russia—where population numbers have stabilized or even declined (i.e., where population growth rates have turned negative). In fact, if suddenly, by some work of magic, all adult couples on Earth were able to limit their reproduction to just one child, the global human population would sink back to below one billion within two hundred years (i.e., in the same span of time it has increased from one to almost eight billion).

The Instructive Case of Kerala

It has long been noted that people in LDCs tend to have more children than people in more developed countries (see box), but there are fascinating exceptions to this pattern. Take the state of Kerala in southern India.

Table 6.1. Quality-of-Life Indicators for Less Developed Countries, Kerala, and the United States

Quality-of-Life Indicators	Less Developed Countries	Kerala State, India	United States
Per capita income ($/year)	400	350	40,000
Life expectancy (years)	58	73	72
Birth rate (N/1000/year)	40	16	14
Literacy (%)	50	96	96

Source: United Nations Population Fund (UNFPA), "Population Dynamics in the LDCS: Challenges and Opportunities for Development and Poverty Reduction," www.unfpa.org; http://en.wikipedia.org/wiki/List_of_countries_by_population_growth_rate; http://en.wikipedia.org/wiki/Kerala#Human_Development_Index.

The citizens of Kerala are poor—their per capita income mirrors that of LDCs, but that's where the similarities end. Indeed, in spite of its meager per capita income, Kerala looks almost identical to the United States in terms of key quality-of-life indicators such as life expectancy, birthrate, and literacy (table 6.1).[3]

Why does Kerala seem to defy the rules of poverty? Science writer Bill McKibben, who traveled to Kerala to unpack this paradox, frames it this way:

> Kerala undercuts maxims about the world we consider almost intuitive: Rich people are healthier, rich people live longer, rich people have more opportunity for education, rich people have fewer children. We "know" all these things to be true—and yet here [in Kerala] is a counter case. . . . It's as if someone demonstrated in a lab that flame didn't necessarily need oxygen, or that water could freeze at 60 degrees. It demands a new chemistry to explain it, a whole new science.[4]

There are a variety of factors that explain the paradox of Kerala, the most significant of which may be the respect accorded to women in Kerala. Typically in LDCs women are subjected to systemic discrimination insofar as they routinely receive less education, own less land, have less political power, get paid less money, work longer hours, and exercise less control over their reproductive lives than their male counterparts. But according to McKibben and others, Kerala's "new chemistry" has ingredients that are rarely seen in other cash-poor places—namely: equal rights for women, effective government, redistribution of wealth, and full access to education and health care.[5]

The educational opportunities afforded to all of Kerala's citizens have been especially important for women because educated women are better able to take charge of their lives. As a reflection of this, the average woman in Kerala doesn't marry until twenty-two years of age, versus eighteen years in the rest of India. And as a further testimony to the regard accorded to women, Kerala is the only place in all of Asia where young girls outnumber young boys—a strong indication that gender is not a factor in abortions.[6]

USING FRESH QUESTIONS TO REFRAME OLD PROBLEMS

We frame a picture to give it definition. If we choose the wrong size frame, the picture doesn't look right. In like manner, we often use questions to frame problems. If our questions are too small—for example, if they emerge from rigid or confining thought patterns—we might not be able to solve the problem before us. Ellen Langer, in her book *Mindfulness*, provides a nifty example of how narrow, conditioned thinking can cripple problem solving:

Imagine that it's two o'clock in the morning. Your doorbell rings; you get up, startled, and make your way downstairs. You open the door and see a man standing before you. He wears two diamond rings and a fur coat and there's a Rolls Royce behind him. He's sorry to wake you at this ridiculous hour he tells you, but he's in the middle of a scavenger hunt. . . . He needs a piece of wood about three feet by seven feet. Can you help him? In order to make it worthwhile, he'll give you $10,000. You believe him. He's obviously rich. And so you say to yourself, how in the world can I get this piece of wood for him? You think of the lumberyard; you don't know who owns the lumberyard; in fact you're not even sure where the lumberyard is. It would be closed at two o'clock in the morning anyway. You struggle but you can't come up with anything. Reluctantly, you tell him, "Gee, I'm sorry."

The next day, when passing a construction site near a friend's house, you see a piece of wood that's just about the right size, three feet by seven feet—a door. You could have just taken a door [from your house] off its hinges and given it to him for $10,000. Why on earth, you say to yourself, didn't it occur to you to do that? It didn't occur to you because yesterday your door was not a piece of wood. The seven-by-three-foot piece of wood was hidden from you, stuck in the category called "door."[7]

Just like the door story above, when we don't stop to question the categories we take for granted, we become stuck in conditioned ways of thinking. For example, the population problem was, for many years, framed in terms of poverty: Poor people were simply having too many children and most of those children were now surviving because of advances in modern medicine! The solution was birth control—plain and simple. Fortunately, nowadays, population growth is coming to be understood in the context of a deeper set of issues rooted in systematic discrimination against women in social, educational, political, and economic realms. With this shift in perception, the question is no longer how can women be kept from having more children (a narrow and demeaning frame) but, instead, how can women be empowered in ways that enhance their freedom and autonomy? When this happens, as we see in the case of Kerala, women, of their own accord, opt to have fewer children.

The Population-Consumption Link

It would be easy to conclude, given the foregoing, that the ecological havoc being wrought on Earth by humankind is primarily the result of humanity's surging population—that it's a population problem, plain and simple—and that if we can simply stabilize our population, everything will be OK. What do you think? Would that do it?

Before answering, consider that the dramatic rise in human numbers in recent times has been mirrored by an even sharper rise in the production and consumption of toasters, fertilizers, plastics, cars, paper, guns, cell phones, computers, shoes, not to mention hundreds of thousands of other items. This growth, in turn, has led to dramatic increases in the by-products of both production and consumption, including solid waste, hazardous waste, freshwater and ocean pollution, greenhouse gases, as well as precipitous declines in forest cover, ocean fisheries, topsoil, global biodiversity, and more (figure 6.1).[8]

The Consumer Lifestyle

Have you ever noticed how U.S. media venues often refer to you and me not as citizens or Americans or even people, but—as "consumers"? In some ways "consumer" is an apt descriptor. If you are reading this book, you probably fit into the "consumer" category.

Overall, the rate at which humankind, as a whole, is now consuming Earth's resources is much greater than the rate of human population growth. This means that, even if population size was to stabilize today, our

HOMO COLOSSUS

The consumer lifestyle results in the consumption of prodigious amounts of energy. When humans lived as hunters and gatherers, each person consumed about 2,500 calories of energy each day to survive, almost all of it in the form of food. This is roughly equivalent to the daily energy intake of a dolphin. But these days, those of us in the *consumer* category (i.e., in the "developed" world) use, on average, about 180,000 calories each day, most of it in the form of fossil fuels. This is akin to the daily caloric intake of a sixty-ton sperm whale. For consumers, it is as if our stomachs have expanded seventy-fold. We have become giants.[9] Referring to our colossal consumption, the ecologist Eugene Odum suggested that our species name should be changed from *Homo sapiens* to *Homo colossus*.[10]

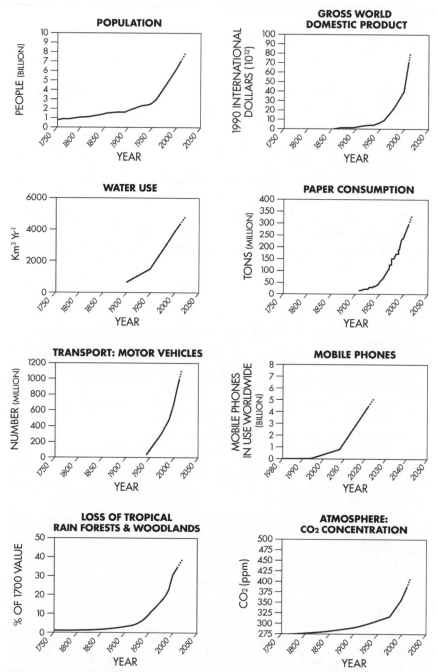

Figure 6.1. The extraordinary rise in human population is mirrored in the dramatic increase in consumption and environmental impacts.

collective impact on Earth will still grow because each of us, on average, is consuming more and more stuff each year. For example, the average American in the first decades of the twenty-first century was consuming three times as much as he or she did in 1960.[11]

All the stuff that we consume, of course, comes from somewhere and that "somewhere" is Earth. In the year 1980, the world's population used an estimated forty billion metric tons of Earth materials, consisting of metals, fossil fuels, biomass, and minerals. Three decades later that amount had increased by almost 50 percent to some sixty billion tons. This doesn't include the tens of billions of tons of waste generated in production processes.

These are big numbers, so let's consider just the United States in the year 2000. That year, total U.S. natural resource consumption, including waste and imports, was 18 billion tons. This means that each American consumed, on average, 132,000 pounds of resource-related materials consisting of oil, sand, grain, iron ore, coal, and wood. Even that number is hard to grasp, so try this: In 2000 each American consumed an eye-popping 362 pounds of natural resources each day—fifteen pounds an hour! And mind you, that was back in 2000.[12] Since then, material consumption has continued to rise.

Another palpable measure of the increase in U.S. consumption is the rise of storage lockers. The storage industry was essentially nonexistent prior to 1970. Since then it has grown so precipitously that its structures now extend over more than seventy square miles—equivalent to twenty-one square feet of storage space for every household in the United States.[13]

But the bulk of what Americans purchase isn't stored; rather, it is used for a short time and then discarded. For example, just in one year (2007) Americans discarded 140 million cell phones, each containing hazardous substances such as lead, arsenic, and cadmium that are associated with some of the same cancers and reproductive abnormalities discussed in the previous chapter.[14] Even though recycling options are now available, the majority of used cell phones still end up in landfills where their toxins slowly leach into Earth's soil and water. And it is much the same story—easy come, easy go—with the other electronic devices that we now surround ourselves with.

Why We Consume

Our very biology predisposes us to be materially oriented. After all, we humans are, by nature, a curious lot, readily attracted to what is novel. And we are intensely social. This means that if we like some new gizmo, we will spread the word to others. It also means that we are highly aware of what those around us are doing and where we stand, in terms of social status, compared to others. Indeed, a large body of research reveals that our motivation (largely unconscious) for buying the hottest new thing on the market is to enhance our status and/or social position. In effect, our possessions contribute, at least in some measure, to our sense of meaning and identity.[15]

Add to this the fact that, as humans, we are prone to habituation. For example, just now I took a break from writing to purchase a cup of java. I didn't need to do this. So why did I? Well, it has become a habit—something I do every afternoon—and this habit of mine is hard for me to break. It is no different, really, with things like electricity, central heating, air conditioning, microwaves, iPods, and computers. Before these things existed, humans didn't need them. Now, most of us are habituated to them and can't imagine life without them.

In addition to novelty and *keeping up with the Joneses* and habituation, addiction is a final factor driving the hyper-consumerism of the modern age. We are all addicts to varying degrees—addicted to our thinking, our imagined needs, our momentary yearnings. An addict's craving is often characterized as an *empty hole* that can never be filled by the addictive activity because that activity (shopping, eating, gambling, Facebook, video games) is merely a substitute for what is really necessary to fill the addict's *empty hole* once and for all. And what is that? Stay tuned for part III.

CRAVE → ACQUIRE → RELIEF. . . .
CRAVE → ACQUIRE → RELIEF. . . .

Have you ever wanted something really badly? Maybe it was a brand-new BMW! That BMW was all you could think about; you were obsessed with owning it. And so you worked and saved and finally, after three years, you had in your very hands the keys to the object of your dreams, a silver BMW. You were elated, giddy, ecstatic. But then, a short time later, your joy had passed. The spike in happiness that the BMW had brought to you was short-lived—it filled a *hole* inside you—but not for long. This prompts the question: Was it actually having that BMW finally in your grasp that brought you that momentary spike in satisfaction, or did your happiness come more because, for a short time, you were living in a state that was free of craving? Indeed, acquiring stuff can become an addiction: Crave → Acquire → Relief; and, then, after a short while, craving returns and the cycle repeats. It is as if the getting—whether it is a bottle of beer or a new outfit—can never bring lasting relief, because it is merely a poor substitute for what we are truly needing.

Manufacturing Needs—The Case of Bottled Water

Although, as humans, we are biologically predisposed to be interested in snazzy stuff, the current *hyper-consumer* lifestyle didn't just happen as a natural result of our biology. It is a cultural phenomenon, as much as anything—the result of the particular economic system that we have created. In this vein, consider that marketers in the United States now spend in excess of $200 billion a year (equivalent to $600 per capita) to convince Americans that they need what they, the marketers, have to sell. No surprise then that many people find themselves buying products that they didn't even know they needed. For example, today, more than half of all Americans drink bottled water. In other words, they pay top dollar for something that is available for a fraction of a penny per glass from the tap. Most of what people pay for when they buy bottled water is the container and the cost of transport. After all, bottled water usually comes from hundreds, even thousands, of miles away. All this transport combined with the manufacture of the plastic bottles requires fossil fuels. It is estimated that a half cup of oil is required to produce a one-liter plastic water bottle. Ironically, approximately two liters of fresh water are also used in the manufacture of each one-liter plastic bottle.[16] Here is a further irony: It turns out that 25 percent, or more, of the bottled water that is sold comes from the tap, with a few minerals added.[17]

Information like this calls into question the health and safety claims made by bottled water purveyors. Indeed, some years back the Natural Resources Defense Council tested the quality of 103 brands of bottled water and found that one-third of the brands contained significant impurities (e.g., levels of chemical or bacterial contaminants exceeding those allowed under state or industry standards). The upshot: In general, there is no guarantee that bottled water is any safer than tap water. In fact, the National Resources Defense Council study reported that tap water is subject to more stringent regulations than is the case for bottled water.[18]

And the story doesn't end there. What happens to all those plastic bottles after the water is consumed? The majority are discarded and end up in landfills where they may require thousands of years to biodegrade. As for the ones that get recycled, there is a good chance these days that they will be exported as far away as China for resource recovery.[19]

I have singled out bottled water for illustrative purposes but there are thousands upon thousands of other products and gizmos put on the market each year, each one whispering: "Buy me!" "You need me!" Simply visit a nearby mall or a superstore and cruise up and down the aisles perusing the

---

COLLEGES SAYING "NO" TO BOTTLED WATER

Given the questionable wisdom of bottled water as a commodity, some student activists have prodded their colleges and universities to institute policies banning the sale of bottled water. As one Dartmouth student observed, "The product just doesn't make common sense. Companies are taking something that is freely accessible to everyone . . . packaging it in a non-reusable container, and then selling it under the pretense that it is somehow better than tap water."[20]

---

myriad offerings. As you go along, ask yourself, could I live without this product? Will it truly enhance my well-being? Will it bring me genuine happiness? When I did this recently at my local big-box store I was amazed to realize that much of what I was seeing didn't even exist as a product category when I was growing up. In other words, I was being told that I needed things that until recently have not even existed!

In sum, the so-called environmental crisis is not simply the result of rapid human population growth. Consumption, it turns out, is the more critical factor. Each item that we buy—whether it's a pack of gum, a car, or an iPod—comes from Earth and goes back to Earth, leaving a trail of impact in its creation, use, and disposal.

FOUNDATION 6.2: FIRST PERSON ECOLOGY: OUR ECOLOGICAL FOOTPRINT

We live in a remarkable historical epoch: population rising, production growing, consumption climbing, technology expanding. But can it continue? Does Planet Earth have enough wealth for all of us to consume as we see fit? This is the question we turn to now.

Fortunately, there is a metric, called "the ecological footprint," that has been explicitly devised to address this question. You have, no doubt, heard of it. If you need reminding—a person's ecological footprint is the area of productive land and sea necessary to produce all the things that s/he consumes each year plus the space necessary to absorb the waste that s/he produces.

The footprint concept is based on two straightforward assumptions: (1) it is possible to accurately estimate both the stuff each of us consumes and the waste that each of us produces; and (2) our consumption and waste can be expressed in terms of Earth-area equivalents.

In practice, calculating ecological footprints is an accounting exercise: Consider a hamburger that you might order at a fast-food joint (ignoring, for the time being, the burger's bun and condiments). The beef patty came from a steer. During its early life, the steer required land to graze on; and later, when it was being fattened in a feedlot, land was also needed to grow the steer's feed. When it was full grown, the steer was killed and processed at a meatpacking plant, and this necessitated space for the facility and energy to run the plant's operations. Next, the processed beef was loaded on a refrigerated truck and transported to burger outlets, entailing more expenditure of materials and energy. Each food outlet offering burgers also has its footprint, owing to the energy and materials used in its construction and operations, as well as the actual space occupied by the establishment. By piecing together the production history of the burger in this way it is possible to estimate the amount of Earth space required to produce it.

Food is one part of a person's footprint; other components include clothing, housing, furniture, appliances, and electronic devices, not to mention one's energy needs for heating, cooling, electricity, and transportation.[21]

The Power of Footprinting

With footprinting humankind is now able, for the first time, to determine the capacity of a region or nation or planet to support its human population. The prospects, good or grim, depend on the extent of productive acreage relative to the number of people and the amount that those people are consuming.

So how much productive acreage is there on Earth? The answer: 32.3 billion acres, including crop, pasture, and forest land, as well as productive ocean surface. That's the size of the *pie*. But wait a minute! What about the millions of other species that are here on Earth with us? They also need Earth's productive land and sea to flourish. Taking this into account, the World Commission on Environment and Development recommended that a minimum of 12 percent of Earth's productive surface (or 3.9 billion acres) be preserved in its natural state. Setting this amount aside leaves humans with a 28.4 billion acre pie to meet our needs (32.3 − 3.9 = 28.4 billion acres). The amount of this *pie* that an individual person *eats* (i.e., requires to support his/her lifestyle) is that person's individual ecological footprint. What's your guess as to your footprint size? One acre? Ten? A hundred? A thousand? You can get a rough estimate by going to www.earthday.org/footprint-calculator and filling out a footprint quiz.

Table 6.2. Productive Acreage on Earth (in billions of acres)

Cropland	3.7
Pasture land	8.5
Forest	12.8
Productive ocean surface	7.3
Total	32.3
Adjusted total	28.4[a]

[a]Assumes setting aside 12 percent of productive land and sea as parks and preserves.

Insofar as consumption levels vary from country to country, it should come as no surprise that footprint sizes also vary by country. For example, the average ecological footprint of U.S. citizens is twenty acres. That's about twice the size of the footprint of the average European and eight times more than the average citizen of India.[22] Meanwhile, within any given country, the richer the person, the higher his/her footprint tends to be since as consumption generally increases in accord with income. In the United States, for example, someone earning $200,000 per year would likely have a footprint greater than fifty acres, while someone taking home only $20,000 per year would, in all likelihood, have a footprint less than ten acres. The same goes for education: the better educated a person is (i.e., the more degrees s/he holds), the higher that person's footprint will tend to be. The inescapable conclusion is that it is the prosperous and best educated among us who often exact the greatest toll on Earth.

Footprinting can also be used to assess a country's vulnerability to economic threats. For example, knowing a nation's population size and its average per capita footprint, we can ask if that country has enough productive land/sea within its boundaries to support its population at current footprint size. Some countries (e.g., Brazil, Australia) have huge amounts of productive area and relatively low populations. Such countries, given their ecological assets, have better prospects for withstanding a prolonged economic crisis. By contrast, places like Singapore, with sparse ecological assets, have to rely on productive acreage outside their borders to supply their people's needs. Overall, out of the 153 countries where footprint assessments have been completed, 107 don't have enough productive acreage within their borders to supply their needs.[23] This is true of the United States. In fact, more than half of America's aggregate footprint requirement is imported from elsewhere.[24] In other words, the United States is a debtor nation in ecological terms—that is, it doesn't come close to supplying its citizens' enormous material consumption demands with the productive land and sea within its borders.

The U.S. situation, it turns out, reflects that of the planet as a whole. Humanity's global footprint exceeds the total productive acreage of Earth by 50 percent. Such an overshoot means that it takes about eighteen months for Earth to regenerate the resources that humankind consumes in twelve months. In other words: We would need 1.5 Earths to sustain global consumption at current rates.[25]

But, wait a minute; how this can be? Why are humans still around if we now require 1.5 Earths to supply our needs and there is only one Earth available to meet our needs? To deconstruct this paradox, consider, first, that Earth has enormous stocks of natural capital (see chapter 5) in the form of soil, water, forests, fish, minerals, and more. Because of this abundance, humans have been able to get away—at least for a while—with drawing down Earth's natural capital stocks more quickly than these stocks have been replenishing themselves. However, if humans were to persist in this overdraw, they would eventually exhaust Earth's natural capital and ecological breakdown would ensue. It turns out that this, in effect, is the story of the last two hundred years. Humankind has been taking too much too fast and, as a result, we have been depleting the water in Earth's aquifers, the soil in Earth's crop and pasture lands, the wood in Earth's forests, and the fishes in Earth's oceans. Simultaneously, we have been putting too much waste too fast into the biosphere, and in the process toxifying Earth's soil and water and air . . . all the while undermining the stability of Earth's climate.

Though you may be just now learning about this emerging ecological crisis, it was actually foreseen back in the early 1970s when scientists at MIT provided evidence that continued unbridled growth in population, consumption, and waste generation would, in coming decades, lead to the degradation of Earth and with it a sharp decline in human well-being. The analysis behind this prediction was laid out in a slender volume, *The Limits to Growth*.[26] Though those early researchers weren't right in every respect, they were prescient in recognizing that "business as usual" could only go on for so long. Indeed, there is growing evidence that we are headed toward a breakdown (some even predict collapse) in our biological support system. Ecologist Jared Diamond, who has made a point of studying the breakdown of past civilizations, writes: "One of the main lessons to be learned from the collapses of the Maya, Anasazi, Easter Islanders, and those other past societies, is that a society's deep decline may begin only a decade or two *after* the society reaches its peak numbers, wealth and power."[27] This really isn't so surprising insofar as peak wealth can be expected to correlate with peak impact on a society's biological support functions.

STEPPING BACK TO SEE THE BIG PICTURE

> Insanity is doing the same thing over and over again and expecting a different result.[28]

Insofar as the ecological footprint metric can be used to calculate the footprint size of individuals living in vastly different socioeconomic situations, it provides a lever for considering some *big picture* ethical questions. For example, who, in your view, bears more responsibility for the beleaguered condition of Earth—a couple from India that has six children or an American couple with only two kids? Certainly, six children seems a bit excessive—even irresponsible—but look again, this time through the lens of footprinting. The average citizen of India has a footprint size of only 2.5 acres. Thus, that large Indian family has an aggregate footprint of 20 acres ([2 adults + 6 children] × 2.5 acre footprint each). This is four times less than the small American family with its aggregate 80-acre footprint ([2 adults + 2 children] × 20-acre footprint each). From the perspective of Earth, of course, what matters is not family size, but the piece of Earth's pie that each family commandeers. Hence, people living in highly developed countries like the United States appear to have a more serious "population problem"—that is, they exert a far more significant impact on Earth—than people living in LDCs. This is not about assigning blame so much as being honest with ourselves and each other.

Here's a second example of how footprinting can serve as a lever, providing insight into important ethical concerns. It starts with a simple question: How much productive acreage would be available for each human inhabitant of Earth if we all received the same share? We can calculate this number by simply dividing the productive surface area of Earth by the human population size. The answer comes to four acres; this would be a *fair-earth share*, assuming Earth's resources were divided equally among all human beings.

Now, consider that the current U.S. per capita footprint of 20 acres is five times more than the current *fair-earth share*. Indeed, if everybody on Earth had a U.S.-sized footprint, Earth could only support 1.4 billion people (28.4 billion productive acres/20 acres per person). Another way of saying this is that approximately five Earths would be necessary to support the world's current population at U.S. standards. Calculations like this reveal the fallacy of imagining that—given current technologies and consumption patterns—all the people now on Earth could ever come close to enacting U.S. lifestyles.

IMAGINE THE CHINESE PEOPLE WITH U.S.-SIZED FOOTPRINTS

In recent decades, China's economy has been growing at a phenomenal 10 percent per year. Applying Rule-70, this means China's economy has been doubling every seven years (70/10 = 7 years). If this extraordinary economic growth rate were to continue, China's 1.3 billion people would achieve the same per capita wealth enjoyed by U.S. citizens in less than two decades. Now, let's just imagine that this happens and that the Chinese people use their added wealth to increase their consumption of things like grain, oil, paper, and cars to current U.S. levels. Wonder what that would mean? Here's what: They would be consuming 1,350 million tons of grain per year (That's 60 percent of the entire world grain harvest in 2011!). Oil? They would be consuming 99 million barrels of oil a day (That's more oil than the entire world consumes today!). Paper? They'd be using almost twice as much as the world consumes today. Cars? They'd have more than a billion on the road—more than the total number of cars, worldwide, in 2012. And that's just China. Nearby India, with its population of more than one billion people, has an economy growing at almost 8 percent per year.[29] These sobering statistics remind us that there are limits to growth on a finite planet.

Yes, all of this footprinting can lead to some downright depressing conclusions, but there is a bright side. Footprinting finally provides humankind with a tool for candidly assessing its situation. Indeed, footprint assessments are like bank statements, and bank statements are great tools—even when they tell us that our finances are in disarray—because they provide the information and, hopefully, the motivation required to set things right, provided we can summon the courage to do so.

Now, consider what it would mean, for a U.S. citizen, to live within his/her fair-earth share of resources. Would it even be possible for Americans to reduce their footprint fivefold, from twenty acres to four acres? Recently, I met with some friends to explore this question. We began by looking at the things we consume and then considered footprint reduction strategies for those things. We started with our personal computers. The options we came up with included: (1) Don't bother to regularly replace "old" computers with brand-new ones; instead support a local business to upgrade old computer components as needed; (2) Team up with a friend and share a computer; and (3) Forgo owning a computer altogether and, instead, use those available in public libraries.

We went on to discuss other categories of consumption—e.g., clothing, housing, entertainment, transportation—and, in so doing, came to see that for each category, viable approaches were at least theoretically possible for reducing our footprint size, even as much as fivefold. Later I realized that there was a bigger question undergirding our conversation that day— namely: Is it possible to reduce our ecological footprint and, at the same time, improve (or at least not sacrifice) our happiness?

Based on global surveys by sociologists, human beings all over the world say that what they want most in life—for themselves and those around them—is happiness. They also mention money, but the longing for happiness is far more important than that for money. Based on these findings, researchers at the New Economics Foundation (NEF) have devised an index of progress dubbed "The Happy Planet Index" (HPI). Rather than using conventional measures such as gross national product to assess national well-being, the folks at NEF focus on people's happiness, believing, quite sensibly, that a successful nation is one inhabited by happy people. One exciting outcome of NEF's research is their finding that it is possible for people to have smallish ecological footprints *and* still be very happy. In other words, small footprints and happiness are not mutually exclusive. Costa Rica is a case in point. It ranks number one in NEF's Happy Planet Index insofar as Costa Ricans are very efficient in transforming the Earth resources they consume into human happiness. By contrast, the United States, with a per capita footprint size three times that of Costa Rica, has a less happy populace (according to survey results), and ranks a lowly 105th on the HPI.

For countries (and individuals) interested in improving their HPI score, research reveals that happiness can be enhanced through five simple actions: (1) connecting with others; (2) increasing physical activity; (3) taking notice of one's surroundings; (4) continuing to learn; and (5) giving time and attention to others. The good news here is that none of these happiness enhancers requires consumption or spending money—in other words, none of them costs the Earth anything![30]

In sum, reducing one's footprint need not be framed in terms of undue deprivation and self-sacrifice. Instead, the challenge of living within the ecological limits imposed by Earth could provide us with opportunities to experience genuine fulfillment by escaping the crave→acquire→relief cycle, while at the same time rekindling our fundamental need for relationship, connectedness, and meaning.[31]

WRAP-UP: EXPANDING ECOLOGICAL CONSCIOUSNESS

> We have lived our lives by the assumption that what was good for us would be good for the world. We have been wrong. We must change our lives so that it will be possible to live by the contrary assumption, that what is good for the world will be good for us. . . . And that requires that we make the effort to know the world and learn what is good for it.
>
> —Wendell Berry[32]

In the end, the world that we are creating for ourselves is determined, to a significant degree, by the quality and boldness of our individual and collective questions. The question "Why?" as showcased in this chapter is a particularly powerful lever for resolving problems and discovering the underlying causes of many troubling phenomena. For instance, Toyota gives part of the credit for its high productivity and exemplary quality control to the question "Why?" Every worker on a Toyota assembly line is taught to analyze problems by asking "Why?" over and over again:

This bolt fell off.
Why?
Because the thread is stripped.
Why?
Because it was misaligned with the screw.
Why?

And so on. It turns out that almost every design problem encountered at Toyota can be solved with five whys or less.[33]

In my experience, fearlessly asking "Why?" over and over again (just as young children are prone to do) is an invitation to authenticity—an opportunity to be real with each other and honest with ourselves. For example, imagine yourself, as a parent, having this conversation with your ten-year-old daughter:

Daughter: I just learned in school that there are two billion people in the world who earn less than $2 per day and that many of these people are hungry. Why is that happening?

You: Well, honey, it's just that adults in those places tend to have a lot of kids and there are just too many mouths to feed.

Daughter: But why do they have so many kids?

You: Well, dear, I don't know. Maybe they just don't know any better.

Daughter: But my teacher said that every day thousands of kids in those places die of sickness and hunger. Why is that?

You: [Pause.] Listen, hon, it's just that there is a lot of sickness in those places and sometimes there are bad droughts and not enough food.

Daughter: So, why doesn't anybody do anything to help?

You: Oh, they do. The Red Cross is there helping out in emergencies.

Daughter: Then, why are all those children still dying?

You: I don't know; it's just the way it is. Listen, it's not good for you to dwell on these things.

Daughter: Why not?

You: Oh, it will just make you sad.

Daughter: But why is it bad to feel sad . . . when that's what I feel?

You: Here, have a snack and stop this talk. There is nothing we can do.

Daughter: Why can't we give some of what we have to those children?

You: That would get really complicated. I wouldn't know where to start.

Daughter: Why wouldn't you know where to start?

You: Listen, you just need to accept that there is nothing we can do about this!

Fearlessly digging into the *whys* underneath human suffering and inequity is an invitation to look deeper and deeper into ourselves—especially into our limiting assumptions and self-serving rationales. Indeed, discovering the ways that our limiting beliefs (e.g., *There is nothing I can do about this!*) constrict our minds and deaden our hearts creates discomfort and sadness; but the reward—the expansion and deepening of consciousness—I believe, is worth the effort.

APPLICATIONS AND PRACTICES: LIVING THE QUESTIONS

> I want to beg you, as much as I can, dear sir, to be patient toward all that is unsolved in your heart and try to love the questions themselves like locked rooms and like books that are written in a very foreign tongue.

Do not now seek the answers, which cannot be given you because you would not be able to live them. And the point is to live everything. Live the questions now. Perhaps you will then gradually, without noticing it, live along some distant day into the answer.

—Rainer Maria Rilke[34]

Almost everything we do, as human beings, is based on our beliefs. We believe that it is necessary to eat three meals a day and so that is what we do. Ditto for eight hours of sleep a night and six glasses of water per day. But, of course, these, and many similar statements, are just assumptions, beliefs that have become so completely kneaded into our psyches that we mistake them for essential truths. But each one can and should be questioned.

Using Questions to Unmask Assumptions

The more our lives are cordoned off by unquestioned assumptions, the more we run the risk of sacrificing our freedom, spontaneity, creativity, and overall aliveness. This is especially true in the sociopolitical sphere. For example, take a minute to carefully consider the following statement: Technology will solve all of our problems. Do you agree? Is this true? Why? How do you know? What if it's false? Read it again, this time with the awareness that it is simply an assumption worthy of your scrutiny.

Social commentator Jerry Mander suggests that new technologies sometimes create more problems than they solve. So it is that Mander recommends that we judge all new technologies as "guilty until proven innocent." By way of illustration, consider this vignette from Mander's book, *In the Absence of the Sacred*:

In the early 1900s the car was portrayed as a harbinger of personal freedom and democracy: private transportation that was fast, clean (no mud or manure), and independent. But what if the public . . . had been told that the car would bring with it the modern concrete city? Or that the car would contribute to cancer-causing air pollution, to noise, to solid waste problems, and to the rapid depletion of the world's resources? What if the public had been made aware that a nation of private car owners would require the virtual repaving of the entire landscape, at public cost, so that eventually automobile sounds would be heard even in wilderness areas? What if it had been realized that the private car would only be manufactured by a small number of giant corporations, leading to their acquiring tremendous economic and political power? That these corporations would create a new mode of mass production—the assembly line—which in turn would [contribute to] worker alienation, injury, drug abuse, and alcoholism? That these corporations might conspire to eliminate other means of popular transportation, including trains? That the automobile would facilitate suburban growth, and its impact on landscapes? What if there had been an appreciation of the psychological results of the privatization of travel and the modern experience of isolation? What if the public had been forewarned of the unprecedented need for oil that the private car would create? What if the world had known that, because of cars, horrible wars would be fought over oil supplies?

Would a public informed of these factors have decided to proceed with developing the private automobile? Would the public have thought it a good thing? . . . I really cannot guess whether a public so well informed, and given a chance to vote, would have voted against cars. Perhaps not. But the public was NOT so informed. There was never any vote, nor any real debate. And now, only [a few] generations later, we live in a world utterly made over to accommodate the demands and domination of one technology.[35]

Mander, in effect, is exhorting us to ask questions—inconvenient questions—questions that cause us to stop and really think about the seldom acknowledged downside of many of our technologies. In this vein, I am struck by the strong connection in my life between car ownership and personal consumption. When I have a car in my driveway, I can (and sometimes do) respond to different shopping impulses—for example, *I think I'll just run out to the store now to buy x or y or z.* It is so easy; a car gives me ready access to a whole world of things, but most of those things, I don't truly need. How do I know? Because when a car isn't available, the urge to buy passes and I find that I get along just fine.

OK, now, it's your turn. I invite you, if you are willing, to use your critical thinking skills to ferret out common assumptions that crimp your thinking—think of them as unquestioned beliefs. For example, maybe you believe that we, as a society, are making progress; or that our economy must continually grow if our society is to flourish; or that competition is the best way to generate innovation; or that you, as a single individual, have no power to change things. Whatever assumptions you come up with, dig in and question them with fervor and courage.

Yes, courage is necessary. After all, questioning assumptions is not easy, especially when it involves confronting cultural dictums. As author Derrick Jensen points out, our culture censors us when we question its sacred icons: "Question Christianity, damned heathen. Question capitalism, pinko liberal. Question democracy, ungrateful wretch. Question science, just plain stupid."[36] When I first read Jensen's statement, I confess to being taken aback. But then I paused and wondered if capitalism is, really, the best economic system we can aspire to. I also wondered about democracy—not its merits—but, rather, whether it is accurate to say that we, in the United States, even live in a democracy anymore? Jensen's statement even prompted to question my own profession—science—wondering about its own inherent biases and limitations. All this leads me to believe that questioning assumptions is a necessary first step to any substantive change, be it personal or cultural.

Using Questions to Distinguish *Wants* from *Needs*

The ability to doggedly question everything is important, especially in a consumer culture where advertisers are intent on convincing people that those things that they may merely *want* are actually essential *needs*.

Here's an exercise that is useful in distinguishing your *wants* from your *needs*. All it requires is a pencil and paper and a measure of curiosity. Begin by drawing a line, lengthwise, down the middle of a sheet of paper. In the left column (Column 1) list all those things that you regard as your essential material necessities—the things that every human being needs to survive (e.g., air to breathe, food to eat, clothes to keep warm, etc.). Then, on the right side (Column 2), list twenty of your most important nonessential material possessions—things you own and use (e.g., phone, toothpaste, bed) but that are not absolutely necessary for your survival.

Now, select five items from Column 2 that you would be able to give up for a while without too much difficulty if you had to. Put the letter *A* next to these five items. Next, select an additional ten items from Column 2 (bringing your total to fifteen) that you could give up but to do so would cause a fair amount of inconvenience and/or stress. Place the letter *B* next to these ten items. Finally, put a *C* next to the remaining five items in Column 2 because to give these up would cause you enormous inconvenience and perhaps even outright suffering.

Now, with your completed table in hand, pick the *C* item that for you feels most like an essential necessity, even though you know that it is really just a powerful want. Once you have made your selection, make a resolution to live without this item for three days. During this time pay close attention to your reactions. In this vein, eco-psychologist Chellis Glendinning maintains that many of us are literally addicted to the modern conveniences that we surround ourselves with.[37] If Glendinning is right, you may experience withdrawal (i.e., pain, upset, confusion, anger, unbalance) when you suddenly deny yourself your daily "fix" of your selected item. To evaluate if this is true, carefully observe your mental, emotional, and spiritual state, day by day, paying special attention to how your perception of things around you changes. Finally—and this is the most important part—note the ways that eliminating this one thing from your life separates you from and/or brings you into fuller relationship with yourself, others, and Earth. It often turns out—based on my experience—that the value of what is learned from this experiment far exceeds the suffering that is endured.

Using Questions to Make Informed Purchasing Decisions

Recently I asked the women in my environmental science class to raise their hand if they were wearing makeup. Most hands went up. I then asked why. I amplified my question saying that I found it odd that adolescent girls apply makeup to try to look older, while older women use makeup in an effort to look younger. What's that all about? After a time, a woman said, "It's like we are not OK, just as we are; makeup advertisements have convinced me that I am never OK just as I am." This woman was able to unveil the central strategy of many advertisements—to play on people's insecurities, convincing us of our insufficiency while promising that a certain product will alleviate our angst. It's brilliant psychology and sells a lot of product, but leaves us bereft of the one thing that would genuinely bring us satisfaction—namely: full-hearted acceptance of ourselves, just as we are. But we need not be victims. Indeed, having an awareness of how advertising can sometimes prey on our insecurities is a step toward expanding consciousness. Beyond this we can use our intellect and our core values as filters when making purchasing decisions.

This requires that we stop before making a purchase to determine if the object of our desire squares with our values. By way of example, imagine that you have just purchased your first home and when the fall comes you notice that your lawn is covered with leaves. So, you head out to the hardware store to buy a rake. Once there you locate a sturdy, top-of-the-line bamboo rake for $35. It seems like a lot of money to spend and, as you stand there trying to decide, you notice that leaf blowers are on sale, for just $59—not that much more than the bamboo rake. You are tempted to just go for the leaf blower. After all, you like new gadgets; this one's cheap; and it will allow you to get your yard cleaned up in no time. But, then, rather than rushing ahead with your purchase, you pause and ask yourself the question, "What would I be saying 'yes' to if I were to actually buy the rake instead of the leaf blower?" For some people, buying the rake would mean saying *yes* to quiet (instead of noise); to self-reliance (instead of machine reliance); to natural pace (instead of machine-pace); to thrift (instead of unnecessary spending); to simplicity (instead of complexity); to vigorous exercise (instead of labor-saving devices); to clean air (instead of air pollution); to durability (instead of built-in obsolescence). Of course, for others, this discernment process could go in the direction of the leaf blower insofar as it offers innovation, speed, power, and convenience. What about you? What purchases have you made lately and how, if at all, have your values been reflected in those purchases? Again, it all seems to come back to the power and importance of questions. When we pause before buying to ask, "What am I saying *yes* to here?" we have the opportunity to uncover the hidden assumptions embedded in our buying habits, and this, at least in my experience, can expand ecological consciousness.

QUESTIONS FOR REFLECTION

- Over your lifetime what has been your relationship with the question *Why?* and what are some *Why?* questions that arose while reading this chapter?
- What is the size of your ecological footprint (go to: www.earthday.org/footprint-calculator to find out) and what steps could you take to reduce your footprint to a fair-earth-share of four acres?
- How do you understand and experience consumption-related addiction (think: Crave→Acquire→Relief) in your life?
- What is a purchase you made in the last week, and in enacting your purchase what can you infer about your values? That is, in making that purchase, what were you saying "yes" to and what were you saying "no" to?
- There is a tendency—as we learn about all the environmental problems in the world—to throw up our hands and say that there is really nothing we can do. Why do you think people tend to respond in this way? What about you, personally?
- When was the last time you identified an assumption and then proceeded to question it? What did you learn in the process? What's an assumption you are holding that might be ripe for questioning right now?

Part III

HEALING OURSELVES, HEALING EARTH

There are hundreds of ways to kneel and kiss the ground.

—Jelaluddin Rumi[1]

The diagnosis emerging from the preceding three chapters is that Earth is sick. Consider, again, some of the many signs of this ill health: forest fragmentation, migratory bird declines, aquifer depletion, ailing seas, soil degradation, species extinction, bee-colony collapse, climate destabilization, chemical proliferation, endocrine system disruption. These are not isolated problems—they are all connected to the explosive growth in human population and consumption and the concordant proliferation of waste being channeled to Earth's soil, water, and atmosphere.

Just as a human being who is experiencing stress slowly sickens, so it is that Earth now sickens, suffering under the weight of humankind's expanding ecological footprint. All of this is occurring, in no small part, because our old story—grounded in speciesism, control, rampant consumption, hyper-individualism, and incessant growth—is no longer working. Indeed, our times are tumultuous precisely because our old story is dying. Taking all of this into account, Eckhart Tolle, in his book *The Power of Now,* writes:

> At this moment in history, humans are a dangerously insane and very sick species. That's not a judgment. It's a fact. As a testament to "insanity," humans killed a hundred million fellow humans in the twentieth century alone. No other species violates itself on such a grand scale. Only people who are in a deeply negative state, who feel very bad indeed, would create such a reality as a reflection of how they feel.[2]

Tolle, in effect, is suggesting that the world that humankind has created—pocked as it is with pollution, abuse, and strife—is a direct reflection of our collective inner world. Certainly, a nerve is touched when we summon the courage to open our eyes and look straight-on at the suffering, separation, and violence now being inflicted, not just on humans, but on the whole family of life. Facing the pain of our times—allowing ourselves to feel—is a necessary prelude to waking up.

Already, throughout the world—in households, in neighborhoods, in businesses, in churches, in schools, in the countryside, and in the city—an awakening process is under way. Often it begins by asking simple questions such as:

- Is this it?
- Is this what it means to be alive, to be a fully expressed human being?
- Am I living the soulful life of purpose and meaning I was created to live?

The answer is often, "No, this is not what I am here for." This "No" is important. Saying "No, I don't believe this. . . . No, I won't do this anymore. . . ." is a prelude to creating a world that deserves our "Yes."

In the final two chapters of this book I describe the disabling stories and belief traps that have ensnared us, thereby leading to ecological havoc (chapter 7). This is followed by an elaboration in chapter 8 of the *new story* of kinship and community and deeper purpose now emerging all around us.

7

THE OLD STORY

Economism and Separation

Those who do not have power over the story that dominates their lives, the power to retell it, rethink, deconstruct it, joke about it, and change it as times change, truly are powerless, because they cannot think new thoughts.

—Salman Rushdie[1]

Residents of southern India have a clever way of capturing monkeys. They drill a hole in a coconut, place rice inside, and then secure the coconut to a tree with a chain. Here is the clever part: The hole in the coconut is just large enough for a monkey's hand to fit inside, but too small to remove once the hand is filled with rice. Instead of letting go of the rice, monkeys often hold on, greedily, only to be captured in a net by villagers.

Just as monkeys are captured because of their refusal to let go, it may be that we humans are now in a quandary because we are tenaciously holding on to—unwilling to let go of—counterproductive ways of thinking and acting.

An essential first step in addressing the environmental and social challenges of our times is to see clearly that we live today—as has been true of humans down through the ages—in a culture-based story that significantly shapes our thoughts and our consequent behaviors. Most of us (and I fully include myself here) are often not aware that we are playing parts in our culture's grand story. That's the way it is with cultural stories that are effective. We inhabit them so fully and seamlessly that they escape our notice.

My goal in this chapter is to, as best as I am able, unmask and shed light on our cultural story. If our story were working, there would be no need to do this, but there is a growing consensus, rising from many quarters, that it is no longer working very well.[2]

FOUNDATION 7.1: A STORY ROOTED IN SEPARATION?

Beginning in the mid-1990s students began to recommend Daniel Quinn's book *Ishmael* to me. Having now read *Ishmael*, I see why it has attracted my students' attention. This book offers meaning in a time of confusion by seeking to explain how humanity's fractured relationship with Earth is a product of Western culture's overarching story.

According to Quinn, a culture's story is a living mythology that explains how things came to be and how we are to act. Quinn offers Nazi Germany as a vivid example of how a people's story can have disastrous consequences. Hitler gave the German people a story that told how the Aryan race had been discriminated against and abused by mongrel races over history. His story went on to describe how Aryans would rise up and wreak vengeance on their oppressors and then assume their rightful place as the master of all races. Many Germans didn't see Hitler's story as a misguided attempt at making meaning; instead, they embraced it as their destiny. Similarly, many of the citizens of Greece in the time of Homer probably didn't regard their stories as Greek mythology, but simply as the way things were.[3] Our times are no different:

> Like the people of Nazi Germany, [we, too,] are the captives of a story. Of course, we don't even think that there is a story for us. This is simply because the story is so ingrained that we have ceased to recognize it as a story. Every one knows it by heart by the time they are six or seven. Black and white, male and female, rich and poor, Christian and Jew, American and Russian . . . we all hear it. And we hear it incessantly, because every medium of propaganda, every medium of education pours it out incessantly. . . . It is always there humming away in the background like a distant motor that never stops.[4]

Once we become aware of a culture's story, we see how it is employed to explain and justify behaviors. The Nazi Germany story, for example, provided a justification for the creation of a world that would benefit Aryans. One hundred fifty years ago, white Americans created a similar story of racial superiority to justify slavery.

Today, those in America continue to live within a story. As an American myself, I see how this story was transmitted to me in myriad ways, including through the history and social studies textbooks I studied in school. An underlying theme of those texts was humankind's inexorable march of progress. I learned that my distant ancestors were impoverished and backward but that today, thanks to the development of science and the application of technology, humans were becoming ever more enlightened and advanced. Nowhere was this march toward progress more pronounced, I was told, than in the United States of America, the land of freedom and abundance. For proof of U.S. superiority, I was taught about the virtues of America's democracy, the power of its military, the superiority of its science, the wonders of its agriculture, and so on.

One way to appreciate the power of a culture's particular story is to imagine how it would be for you if you had been born into a society with a very different story. For example, imagine that your culture's story was centered on the belief that this life is merely a preparation or test for the life to come; or that there is nothing you can do to change your situation. Or how about: as you live your life, you are cocreating the universe; or that it is your purpose as a human to give of your unique gifts for the healing of the world. Each of these story themes would clearly lead you to a different stance and response regarding your identity and purpose in the world. As futurist Barbara Hubbard points out: "As we see ourselves, so we become."[5] Herein lies the colossal power of each culture's story.

The overall, big-brush, story of Western civilization is predicated on the assumption that the world—Earth—was made for us. Embedded in this story is the belief that struggle, and even combat, are often the necessary means by which goodness triumphs over evil. So it is that we often speak in terms of life as a struggle or contest. For example, doctors *wage war* against cancer, couples *fight* for child custody, companies *mount campaigns* for a larger share of the market, farmers launch *attacks* on crop pests, citizens *struggle* for their rights, communities *combat* crime, economists seek to *subdue* inflation, and nations *battle* each other for resources. In adopting these *contest* metaphors, our story communicates that winning, domination, and coming out on top are what matters most in life.

Given the foregoing, it comes as no surprise that in our story the emergence of *man* is the capstone event in the history of the cosmos. For example, the Christian Bible teaches that the world was unfinished without man.

It needed a ruler. Man had to subdue the world. Even today, we hear it over and over: Man is conquering the deserts; man is subduing the oceans; man is taming the atom and the human genome; man is gaining mastery over outer space. According to Western mythology—the Western story—man was born to control, manage, and exercise dominion over Earth; this is man's destiny.[6] Implicit in this story is the belief that man is separate—apart from and above—the rest of creation. Indeed, my repeated use of "man" here is meant to underscore the central role of patriarchy in our story.

This notion that man is the climax, the final objective, of the whole cosmic drama of creation is simply a story that Westerners have, for the most part, unconsciously absorbed. But these days it takes a fair measure of naiveté to imagine that the entire cosmic unfolding of the universe, extending back some fourteen billion years, was finally accomplished when *Homo sapiens* appeared on the little planet we call "Earth." After all, since the appearance of man, the universe has continued to expand; new stars have continued to be born; and the principles and processes undergirding biological evolution and speciation have continued to operate, largely unabated, just as if man had never appeared. Even if we were to limit our focus to planet Earth, still, to say that "man is the endpoint of Earth's evolutionary processes" is no less shortsighted than imagining a period far back in time—say when photosynthetic bacteria were the most complex life form on Earth—and thinking that the appearance of these amazing bacteria was the final objective of evolution on Earth. But, of course, the generative creativity of the universe and its "speck," planet Earth, didn't stop with the photosynthetic bacteria, nor has it stopped with us. We are simply a minuscule part of a much, much grander unfolding.[7]

The Age of Separation

Today there is much talk about how human beings are becoming more and more connected to each other via jet travel, global trade, the Internet, social networking, and more. There is certainly a lot to support this claim. And, yet, simultaneously, it seems that we are also becoming more separate from ourselves *and* each other *and* planet Earth. The separation that I am referring to often begins at birth when the modern mother—conditioned by her culture to believe that she (unlike all mothers down through history) no longer knows how to give birth—*separates* from her innate biological intelligence, surrendering her autonomy to medical directives at the time of childbirth. Then, within a year of being born, the newborn is often *separated* from her family and placed in a day care facility. Next comes school, where children are *separated* by age and test scores, and often socioeconomic status, and taught that in order to learn it is necessary to observe at a distance—in short, to *separate from* the subject of interest. Later, as adults, we are expected to exchange our life energy for an abstraction—money—and, in so doing, often engage in work that distances, or *separates*, us from our soul's deepest yearnings. Later still, as we get up in years, it is increasingly likely that we will enter—or be placed in—a retirement home, *separated* from the rest of society. Finally, in our dying, we run the risk of being *separated* from our personal dignity, insofar as we give the medical system permission to enact our death for us.

In some measure, from birth to death, we all swim in these waters of separation, often without even knowing it. No surprise, then, that separation consciousness is woven into most, if not all, of our societal institutions. For example, religion (for all its merits) sometimes contributes to separation by dichotomizing the world—saved versus fallen, my religion versus your religion. Government institutions such as justice, welfare, and education—in subtle and not so subtle ways—separate us from our individual integrity by cultivating dependency, even helplessness. Modern technologies (in addition to their benefits) can create separation by luring us with comfort and efficiency, often at the expense of personal know-how and interdependence. Science, with its emphasis on abstract models and numbers, also creates separation (for better or worse) by objectifying all in its purview.[8]

Don't get me wrong. Every day we each have myriad opportunities to experience connection through our relationships with those in our families, neighborhoods, churches, schools, and workplaces. Yet, it bears noting that we live within a larger system of beliefs and institutions that often engender separation.

Separation is pervasive, in no small part, because those of us born in the West have *inherited*, at birth, a way of thinking called "dualism." It was Aristotle who formally presented dualism as a worldview. Everything in the world, according to Aristotle, was either one way or the other: black or white, true or not true, superior or inferior. Aristotle called this the "Law of the Excluded Middle." It could just as easily have been called the "Law of Separation." Yes, it seems that we have been socialized, to varying degrees, to believe that humans are superior to nature, that rich is superior to poor, that reason is superior to intuition, and so on. Dualism is the seedbed of fundamentalism, that all-too-common thought meme that seduces us into believing that my gender, my school, my values, my religion, my opinions, my politics, my race, my nation is superior to yours.

Though dualistic thinking is comforting (and very, very useful at times!), reality, generally, cannot be crisply parsed into discrete categories like rich versus poor, clean versus dirty, or good versus bad. But, wait a minute! How could anybody argue with the fact that clean isn't superior to dirty? Think again. Some epidemiologists now suspect that our modern fixation on cleanliness—e.g., the excessive use of all manner of sanitizers—can actually contribute to allergies as well as weaken the human immune system, making us more prone to sickness.[9] It seems that having some dirt in our environment is a good thing insofar as it tunes and fortifies our immune system.

OK, but what about "good versus bad"? Good is certainly superior to bad, right? I know that as a child I was socialized to believe that when I did something judged as bad, I (by extension) was bad and that I should always try to be good—that is, to follow social norms. But since then I have learned about the history of the aboriginal Orang Asli people of Malaysia. For them, the idea of labeling a person as "bad" would have been preposterous—akin to punishing a plant that was not growing well. Historically, a person in that culture who was not behaving within the norms of the group was understood not as *bad* but as having forgotten his true nature. The solution was to bring that person more fully into the community circle so that he remembered that he belonged to something greater than himself. Though we are not Orang Asli, we can, if we choose, step outside of the often-isolating effects of good versus bad dichotomization.

In sum, the modern-day repercussions of our separation from life are evident everywhere. Just read the newspapers. We harm and even kill each other everywhere, every day, through acts of negligence and violence. We despoil Earth's air, water, and soil—everywhere, every day—often with little awareness. All of this, it would seem, is the outcome of a society populated by people who experience themselves as more separate *from* than connected *to* themselves, each other, and Earth.[10]

Economism: An Outgrowth of Separation Consciousness?

Imagine that you have been hired to make sense of human culture as it is enacted in the United States at the present time. You set about your task, visiting homes, neighborhoods, shopping centers, churches, workplaces, schools, restaurants, farms, parks, sporting events, and festivals throughout the country. As you listen to what people talk about and note what they do, you discover that many people in this culture, in particular those younger than sixty years old, spend the majority of their time working, shopping, and gazing at screens (e.g., computers, televisions, smart phones), with some socializing (often enacted electronically) mixed in.

Upon closer inspection, you note that these people receive something called "money" for their toil and use their money to acquire a remarkable array of things, most of which they discard after a short while. Overall, you are struck by the busyness and seriousness of these people and by the fact that most of the time they don't appear to be particularly happy.

Later, in trying to make sense of your observations, you encounter the concept of "economism," and then everything begins to click into place. Like many other *isms* (e.g., socialism, dualism, Protestantism, nationalism), economism constitutes a belief system or worldview. Adherents of economism assume (often unconsciously) that their primary purpose in life is to work hard to make money and thereby to achieve the success and the material possessions that will bring security and happiness—that is, they live by the formula: Work→Money→Success→Possessions→Security→Happiness. In this view, success is tightly linked to the accumulation of money and possessions. Economism is so fully a part of American thinking and culture that most of us simply see it as the way things are—end of story.[11]

A measure of economism's effectiveness has been its ability to shape the purpose of many societal institutions and functions. So it is that many Americans now believe that the central purpose of schools is to teach children the skills they will need at work so that U.S. businesses can be

A CRICKET AND SOME COINS

Gerry was walking along a sidewalk at midday in Washington, DC, with a Native American friend from the Bureau of Indian Affairs. People were husslin' and busslin' along and the sounds of honking horns and noisy car engines filled the air. In the middle of all the activity, Gerry's friend stopped and said, "Hey, listen, a cricket!"

"What?" said Gerry.

"Yeah, a cricket," said the Native American man as he peered under a bush and located the cricket.

"Damn," said Gerry. "How did you hear that cricket with all this noise and traffic?"

"Oh," said the native man, "It was the way I was raised . . . what I was taught to listen for." Then he reached into his pocket and pulled out a handful of coins—nickels, quarters, dimes—and dropped them on the sidewalk. When this happened, all the pedestrians nearby stopped . . . to listen.[12]

The moral of this story, of course, is that what we listen for and hear is what most matters to us. These days it is the coin and not the cricket—the world of commerce, not the natural world—that seems to command most people's attention.

competitive in the global economy, and the main purpose of government is to promote policies that will help the U.S. economy flourish.[13] In short, under economism, the needs and values of business—the so-called growth economy—have come to dominate society. Activities that generate a profit or bring a high return on investment are judged as desirable, often irrespective of whether they are wise, wholesome, or even morally defensible.[14] Meanwhile, endeavors that don't generate profits—no matter how worthwhile or noble—are often marginalized . . . or so it would seem.

Given the foregoing, it should come as no surprise that, through the lens of economism, the primary purpose of the natural environment—planet Earth—is to provide resources that will fuel the global economy. Indeed, these days the commoditization of Earth has become so routine that most of us barely register its oddness. But what if someone attempted to pass off human bodies as mere resources to be harvested? After all, each of our bodies is composed of many different chemical elements, and some, like phosphorus, nitrogen, potassium, and iron are valuable commodities. Of course, it would be reprehensible to treat a person's body as a commodity, worth so much per pound. And yet, think about it: This is how our economy treats Earth's forests and grasslands and mountains and oceans and all the life that dwells therein—simply as commodities waiting to be exploited.

Meanwhile, in contrast to the physical Earth, we generally have great respect, even reverence, for the "almighty" dollar, no matter its abstract nature. I make this point to my students by holding a lit match in one hand and a crisp one-dollar bill in the other. As I bring the match closer and closer to the dollar bill, I ask then to register what they are feeling—excited? afraid? curious? Then, I proceed to actually set the dollar bill on fire. Many students react with a mix of shock and anger. Some even say I should be arrested for such a heinous act. It is as if by burning a piece of paper I have blasphemed a sacred religious icon. After taking in their comments, I ask them to consider how burning the dollar bill might have actually benefited Earth—e.g., might one less superfluous purchase be a benefit to Earth insofar as each product we consume has a measurable impact on Earth's environment? My larger motivation for this classroom stunt is to invite students to consider how money—especially organizing our lives around money (as economism conditions us to do)—might separate us from understanding ourselves in deeper and more meaningful ways. None of this is to say that I am in favor of abolishing money—I am not! More on this in the next chapter.

Economism as a Pseudo-Religion?

Economism with its associated beliefs has become so fully integrated into U.S. culture (and increasingly into global culture) that it now appears to act as a kind of pseudo-religion. Yes, strange as it may sound, economism, metaphorically speaking, has the equivalent of deities, high priests, missionaries, places of worship, commandments, and more. I admit that on first blush this metaphor seems a bit far-fetched, but I ask you to join me in playfully exploring it. We can start by considering economism's *gods*—the things in our culture that we regard as all-knowing and all-powerful. For example, here are three candidates:

- The Market—the all-knowing *Father* that we are enjoined to place our faith in.
- Money—the means of earthly salvation that we readily dedicate our lives to acquiring.
- Technology—the supreme source of knowledge and power that we are encouraged to embrace and accept.

Indeed, these three (among other possible candidates) seem to constitute a kind of *trinity* that we are enjoined to place our faith in. It follows that the *high priests* of economism—those who are closest to these deities—would be economists, financial analysts, and technologists.

What about further parallels between economism and religion? For example, might the minions from transnational corporations seeking to convert all nations to free-market capitalism be seen as economism's *missionaries*? And might the outlets of the corporate-controlled media, such as television, where we take in economism's central doctrine—that success and happiness come from stuff—be construed as economism's *churches*? Finally, what about the rules or commandments of economism? Here, too, it seems that, just like a bona fide religion, economism is replete with admonitions:

- Work long hours and you will be judged worthy!
- Shop and you will be happy!
- Invest in the market and you will be rewarded!
- Trust in technology and you will be enlightened!

In sum, it seems that economism has become so institutionalized in America that most people take it for granted as reality—as the way life is and must be! For example, if you happen to be a college student, ask yourself: Why am I in college? What's my goal? My purpose? If your answer centers on the attainment of future financial security, welcome to the crowd! You are an adherent of economism. Like other followers of economism, you, in some measure, have signed on to the *story* that your ultimate success—aye, your deep happiness and sense of meaning and purpose—will be determined, to a very significant degree, by your ability to secure material benefits. And, yet, this is a huge and highly questionable assumption—one that was not generally entertained by young people a short fifty years ago. For example, in the 1960s, fully 80 percent of college freshmen registered their primary college concern as "developing a meaningful philosophy of life"; today only 40 percent have this motivation.[15]

It would be easy to take this last statistic at face value and characterize today's college students as greedy and crass but from where I sit, I think this would be a mistake. For example, recently, when I asked a gathering of college freshmen to reflect on what they most valued in life they spoke of family, friendship, their hometowns, their religion, health, learning, honesty, personal freedom, adventure, and natural beauty. By contrast, when I asked them to characterize the values of their culture, they used words like *competition, consumption, making money, image, posing, greed, convenience, getting ahead.* I was surprised by undertones of anger and cynicism in many respondents. Here were young Americans who had personal values that were, for the most part, generous and life-affirming, living in a culture that they perceived as crass. Though they may have not named it as such, they seemed to be experiencing a fair measure of cognitive dissonance resulting from a separation between their personal values and the perceived values of their culture.

The cognitive dissonance—think separation—that economism engenders in the human psyche is echoed in "Yearning for Balance," a comprehensive analysis of American perspectives on modern life. According to this study, when Americans—irrespective of gender, age, or race—look at the condition of the world today, they come to a similar conclusion:

> Things are seriously out of whack. People describe a society at odds with itself and its own most important values. They see their fellow Americans growing increasingly atomized, selfish, and irresponsible; they worry that our society is losing its moral center.[16]

However you cut it, it seems to come out the same: Economism! This is the cultural story that many of us are now enacting, often unwittingly and with less and less enthusiasm. So, let's dig deeper, now, and examine the toll that economism is taking on both our individual and collective lives.

FOUNDATION 7.2: CONSEQUENCES OF ECONOMISM

Economism—with its formula of: Work→Money→Success→Possessions→ Security→Happiness—is just a story. To be sure, this belief system has served us well in many respects, but it is not without troubling consequences that can no longer be ignored. Let's have a look.

Economism *Can* Separate Us from Our Common Sense

Economism teaches us that it is economic growth that makes the world go round. It chides us to keep working, earning, consuming, and smiling, warning us that no growth = no economy = no happiness.

Indeed, economic growth has become an expectation, a habit, the elixir that soothes our fears and satisfies our wants, or so we have been taught to believe. Until recently, to even suggest that our economy need not continue to grow forever was to risk dismissal as a crank. After all, don't we have to keep growing to survive? Well, actually, no, we don't. Given the enormous advances in labor-saving technologies and worker productivity (e.g., the average American worker today is more than twice as productive compared to a half century ago), we, in the United States, could be working much less but, strangely, we are working more than ever!

Here is a thought experiment that helps to make sense of this seeming paradox. Imagine that, overnight, by virtue of a marvelous new technology, American workers were once again able to double their productivity (as has occurred in recent decades). With all this extra productivity, U.S. employers *could* arrange it so that their employees were working only half as much (e.g., just twenty hours/week) while still maintaining the same overall pay. Just imagine the human creativity, the learning, the community service, the happiness and fulfillment that could be engendered if the American work week were cut in half! However, our economic story does not allow this to happen. Instead, the logic of our reigning economic paradigm, capitalism (the handmaiden of economism), dictates that employers must constantly strive to increase profits. Trapped within the confines of this story, employers have two possible responses to a hypothetical doubling of productivity: (1) Fire half the work force to reduce costs and increase profits; or (2) Keep the work force and increase profits by revving up to double the production of goods. Tragically, the idea of responding to a doubling in worker productivity by halving work hours while, at the same time, maintaining pay levels doesn't enter the rigid calculus of economism. This illustrates how devotion to a story—in this case, an economic story—can override our commonsense notions of human well-being.

This thought experiment also shines light on some of the twisted implications of perpetual economic growth. If increased productivity goes to producing more and more goods, all that extra stuff has to be sold. The problem with this is that there really is a limit to how many tables or refrigerators or cars people can be convinced to buy. But here, too, there are two ways to get around this problem. One it to create new categories of things that people didn't even know that they needed. For example, think of the tens of thousands of new, cute but unnecessary, plastic gadgets put on the market each year that enter our homes, closets, and, eventually, trash cans. The second solution is to take things that people once received for free (like water) or things that they once did for themselves or shared with each other (e.g., child care, cooking, car washing) and to turn these into market products. Both these tactics ensure that the economy keeps growing and Americans keep working and consuming.[17]

Yes, it defies common sense. After all, who wouldn't want to live well and with dignity with a twenty-hour workweek, instead of the current forty-plus hour norm? It's possible—within easy reach, even—but we are entrapped in a story that, at least for now, won't permit it.

But might capitalism be reinvented to accommodate a steady-state economy rather than one based on perpetual growth? This is a good and worthy question.

Economism *Can* Separate Us from a Sustainable Future

What fuels the global market economy? Is it money? Technology? Labor? It turns out that these are all secondary players. The primary player is fossil fuels; the global economy runs on fossil fuels. If you are sitting inside right now, take a look around. Every object you see—this book, the walls, your chair, your clothes, your cell phone . . . all of it was produced using fossil fuels. In short, no fossil fuels means no production = no growth = no economy!

The problem, of course, is that while burning fossil fuels allows us to produce the stuff that we want, it also produces prodigious amounts of carbon dioxide and, with it, climate chaos (see chapter 5). Given this conflict, climate scientists, environmentalists, and financial analysts have been crunching numbers to determine the upper limit that should be put on fossil fuel carbon dioxide emissions between now and 2050 to ensure that we still have a livable planet. The answer—derived from highly sophisticated computer modeling efforts by analysts around the world—is approximately 565 gigatons.[18] Think of this as humankind's carbon dioxide budget between now and 2050. If we can stay within this allotment there is a decent chance that Earth's climate won't unravel. Now, here's the rub: Almost all economic forecasts indicate that carbon emissions will continue to grow at 3 percent per year in coming years. At this rate we'll blow past the 565 gigaton carbon dioxide cap by 2030, twenty years before the allowable time limit of 2050.[19]

But maybe carbon dioxide emissions will be forestalled because there's no longer enough fossil fuel around to permit business-as-usual economic growth? This was the question that a team of London financial analysts and environmentalists addressed in the 2011 Carbon Tracker Initiative.[20] Specifically, these researchers combed through proprietary databases to determine the amount of carbon contained in the *proven* oil, coal, and gas reserves of fossil-fuel-producing nations and energy companies like BP and Exxon. And guess what? They found that present-day fossil fuel reserves, if burned, would release approximately 2,795 gigatons of carbon dioxide into Earth's atmosphere. So, once again, the generally recognized cap on what we can safely burn is 565 gigatons, but the stock that is ready to be burned comes to 2,795 gigatons. Translation: There is five times more oil, coal, and gas waiting to be burned than it is safe to burn, given what we now know.[21]

If governments, oil companies, and policy makers were guided by the Precautionary Principle (see chapter 5), they would opt to leave the majority of our fossil fuel reserves in the ground for the time being, thereby avoiding the risk of climate chaos. But we are a people guided by economism's dictum of constant growth and, given this ideology, it is naive to imagine that big energy producers would keep their proven reserves below ground and out of circulation. After all, this stuff is already economically aboveground—"it's figured into share prices, companies are borrowing money against it. These reserves are their primary assets, the holdings that give companies their value."[22] Rather than cutting back, fossil fuel entrepreneurs—both nations and companies—are going full throttle. Exxon alone is now investing tens of millions of dollars every day in oil and gas discovery efforts. British Petroleum and Shell have shut down their solar divisions, returning their focus to fossil fuels. Indeed, energy companies regard the melting of the Arctic summer ice cover that began in 2007 as a business opportunity. Case in point: In 2012 companies were busy bidding on drilling

THE BIGGEST STORY OF OUR LIVES?

In coming years it is likely that media prognostications on fossil fuel burning and consequent climate change will run the gamut from hopeful to dire. For example, at the time of this writing, there were news reports announcing that carbon dioxide emissions, though still very high, had been inching downward in the United States in recent years (in contrast to the steady planet-wide rise in emissions). The U.S. decline was attributed to a boom in domestic natural gas drilling that created a glut in the U.S. gas market. As a result, power plants were buying up the cheap natural gas and burning it, instead of coal, to produce electricity. On the surface, this appeared to be a good thing for the environment insofar as the burning of natural gas emits less climate-warming carbon dioxide per BTU of energy generated, compared to coal. But, as we have seen, often a plus in one environmental realm shows up as a minus in another. The *minus* in this case has to do with the environmental impacts of natural gas extraction. For example, the recent U.S. natural-gas boom is the result of a new gas-drilling technology that now gives energy companies the capacity to extract previously inaccessible gas deposits located miles below ground. This technology, known as "hydrofracking," involves the injection of huge quantities of highly pressurized fracking fluids (a toxic brew of more than one hundred chemicals dissolved in water) deep into the Earth to fracture the rock, thereby providing channels for gas to migrate to the surface. Hydrofracking, while it gives us access to one valuable resource—natural gas—puts another much more precious resource—clean water—at considerable risk. Hence, in the big picture, this seemingly good news about a decline in U.S. CO_2 emissions was really just a report on business-as-usual economism—that is, a profit-making maneuver (buying up natural gas while it was cheap to ensure continued growth) that resulted in the further liquidation of Earth's nonrenewable resources while running the risk of destroying freshwater resources.[23]

leases made possible by the Arctic melt. This is yet one more example of how—caught in the soulless story of economism—*we* are separating ourselves from a sustainable future.

Economism *Can* Separate Us from Earth

A fundamental precept of modern economics is that as individuals seek to maximize their self-interest, they will, without any conscious intention, benefit society as a whole. This comforting notion implies that humans can consume—that is, take from Earth—at will with no significant negative ramifications. So it is that our economic system lacks effective mechanisms to value and safeguard the very Earth upon which we all depend for our continued survival. There is no price tag put on clean air or pure water or healthy seas or flourishing forests.

To make this oversight more vivid, imagine that you are a CEO and your company owns a hundred thousand acres of mature forest. You acquired this land as an investment and your shareholders expect a good return. Unsure what to do, you consult with a professional forester. She tells you that if you cut your forest, it will take one hundred years to regrow to maturity. Given this time factor and the hundred-thousand-acre extent of your holdings, she advises you to harvest the trees from one thousand acres each year. She tells you that, given today's prices, you will realize a net profit of $1 million when you cut your first one thousand acres. That sounds pretty good, and you like the idea of maintaining a steady income stream by cutting just one thousand acres each year, in perpetuity, all the while maintaining the health and diversity of your forest holdings. But before deciding, you consult with an investment analyst who tells you that your best bet is to clear-cut the entire hundred thousand acres right now and then to invest the hundred million dollars of profit back into the market. He adds that even if you take a conservative track and put your hundred million dollars in treasury bonds, at 3 percent interest, you will still be guaranteed a $3 million annual return—three times the measly $1 million promised by the forester. Yes, the forest will be gone, the soils perhaps degraded, the wildlife partly exterminated, but your stockholders will be happy and you will be fulfilling your fiduciary responsibilities.[24] Economism is like this—it has its own logic that separates its adherents from considering the long-term well-being of Earth's myriad ecosystems.

Economism *Can* Separate Us from Each Other

One of the more worrisome effects of economism is that it can undermine what is most precious and essential to human happiness and well-being—the experience of living in relationship to one another in a genuine community. In this vein, consider that informal face-to-face socializing fell by one-third in the United States from the 1970s to 2000 and that, these days, a surprising number of Americans don't even know their neighbors.[25]

Part of this has to do with the increased time Americans are devoting to work. For example, between 1979 and 2006 the average working person added 180 hours to his/her annual workload. Indeed, the United States leads the industrialized world in terms of the proportion of its population that is employed, the number of days spent working each year, and the hours devoted to work each day. In fact, the average American's workload now exceeds that of Germans, French, and Italians by an average of 270 hours per year. Think of it as Americans working six and a half weeks more each year than their European counterparts.[26] "Even the majority of slaves in the ancient world and serfs during the Middle Ages did not work as hard, as regularly, or as long as we do."[27] And this says nothing regarding the dubious quality, much less necessity, of much modern work. Columnist Ellen Goodman captured this "new normal" when she wrote: "Normal is getting dressed in clothes that you buy for work, driving through traffic in a car that you are still paying for, in order to get to the job that you need so you can pay for the clothes, car and the house that you leave empty all day in order to afford to live in it."[28]

Just as our feverish devotion to work can undermine family and community bonds, the same can be said for the design of our living spaces. Take house size for starters: The floor space of the average U.S. home ballooned from approximately 980 square feet in 1950 to more than 2,300 square feet today. Meanwhile, the average number of people per household has declined from 3.4 in 1950 to roughly 2.5 today. In effect, each American now occupies almost 1,000 square feet of living space; that's more than four 15-square-foot rooms per person—enough space to ensure that household members don't have to interact very much with each other if they don't want to. Add to this the fact that houses today are more spread out. For example, according to the U.S. census, in 1920 the average population density was ten people per acre for cities, towns, and suburbs taken together; today, the population density is approximately four people per acre, meaning that today people are, on average, more than twice as spread out as a century ago. (This in spite of the fact that the U.S. population has more than doubled in the interim!)[29]

It's not hard to see how more space in our homes and more space separating our homes along with more time spent away from our homes translates to fewer opportunities to bump into each other. One upshot is what Bill McKibben calls "hyper-individualism"—the belief that we can meet all our needs on our own and, therefore, don't need each other.

In the end, the independence that economism engenders is an illusion, for though we may imagine ourselves to be independent, in reality, almost all of us are utterly dependent on a vast supply network over which we have almost no control. For example, we depend on a remote, largely invisible, global network to feed us, shelter us, fuel us (with gas and oil), clothe us, and even entertain us.

HYPER-INDIVIDUALISM: A BOON FOR WAL-MART?

It has been documented many times: Wal-Mart comes to town and the local economy—built upon community ties—is jeopardized. Unable to compete with Wal-Mart's low prices, family businesses shut down one by one: local pharmacy, gone; downtown hardware store, gone; mom-and-pop grocery, gone; family sporting goods store, shuttered; quaint dress store, closed. Wal-Mart's low prices eventually drive most, if not all, family operations out of business. And though Wal-Mart promises to bring jobs to town, research reveals that it eliminates a job and a half for every new one it creates, and those new ones are often poorly paid with only minimal benefits.

Ultimately, Wal-Mart's success depends on the creation of hyper-individualistic cultures that accord little value to community ties. As Bill McKibben observes, "For Wal-Mart to prosper, we must think of ourselves as individuals—must think that being individuals is the better deal. But, [if you] think of yourself as a member of a community . . . you'll get the better deal. You'll build a world with some hope of ecological stability, and where the chances increase that you'll be happy."[30]

Economism *Can* Separate Us from Abundance

Consider for a moment all the harmful things occurring on Earth—blasting away whole mountain tops to get at coal, clear-cutting forests often with little attention to sustainable forestry practices, decimating populations of ocean fishes, mindlessly spewing toxins into Earth's air and water, creating climate chaos. Now ask, Why is all this occurring? The answer that usually comes back is "Money." We are told, or tell ourselves, that humans destroy, defile, exploit, and create misery for ourselves and others for the sake of money. But is this really true? Is it really money that leads us to pledge allegiance to an economic system that increasingly seems to be harming our souls, our communities, and our planet? Isn't this a bit simplistic? After all, money is just a medium of exchange, so, by itself, it can't be the real problem.[31]

Well, if it's not money, how about pinning the blame on greed? We've all heard it: "People are just so darn greedy!" There is some truth to this. I know that I sometimes take in and hold on to much more than I need. But this raises the question, what causes me or you or anyone to exhibit greed? Think about times in your life when you acted out of greed. What was going on for you? Maybe it was as simple as taking a bit more than your fair share of ice cream.

I note that my greedy behaviors are invariably preceded by a fear that I am not going to have enough. For example, if I know that there is plenty of ice cream, greed does not arise. So, could it be that the problem isn't with money per se or even with greed, but with the mistaken belief that we live in a world of scarcity? Though most of us have been conditioned to believe in scarcity, the truth is that, even in spite of all the environmental havoc humankind has created over the past two centuries, we still live in a world of abundance.[32] Take food. It is well known that—even in the face of our burgeoning population and our poor stewardship of Earth's soils—there is still the potential to adequately feed all human beings alive today. This same potential exists in the realms of housing, clean water, health care, and education—although if we continue on our present course much longer, we will forfeit this potential.[33]

Sadly, although the capacity exists for everyone to be well provisioned, two billion–plus human beings live in a state of need today. That this occurs is a testament to the power of belief—namely the belief in scarcity and the fear and greed and selfishness that this belief engenders. It is the fear of scarcity that drives the already prosperous to continue accumulating more and more. Even those making six figures often succumb to this fear. For example, ask someone with a big salary how much s/he needs to feel secure and the most frequent answer is "about twice what I now make."

All of this is to suggest that it is the system—the story—in which we live that creates our response, rather than human nature per se. It is the belief in scarcity that prompts those who are well off to still worry about not having enough; the story of scarcity that leads governments to say that lowering carbon emissions would slow growth and is, therefore, unacceptable; the fear of scarcity that leads us to say things like "I can't *afford* to volunteer at the health clinic," thereby undermining our innate tendency for kindness.[34]

In sum, there is way more than enough wealth on Earth to feed, clothe, house, heal, and care for everyone—Everyone! It's not resources that we lack, but relationship. What is scarce is compassion and a sense of our shared humanity.[35]

Economism *Can* Separate Us from Happiness

The tragedy of economism—its dark underbelly—is that it isn't making us any happier. For example, in spite of the fact that Americans have increased their expenditures by almost threefold since the 1950s, they are slightly less happy today than in the 1950s—this according to the National Opinion Research Council. In fact, nowadays, fewer than one-third of Americans count themselves as "very happy." Though we have more stuff than ever before—more education, more music, more electronic devices, more clothes, more televisions, more cars, more spacious houses—we are no happier.

This is not to say that there is no connection between money and well-being. Researchers examining happiness in cash-poor countries like India, Brazil, and the Philippines report that an increase in per capita income from, say, $2,000 to $10,000 per year generally results in a marked increase in reported personal happiness, but beyond the $10,000 figure—that is, beyond where basic needs are met—the correlation between income and happiness dissolves. This result is mirrored, anecdotally, in the fact that the happiness scores of *Forbes* magazine's richest Americans are identical to the Pennsylvania Amish—a people relatively poor in cash but rich in community bonds.[36]

Tragically, Americans assumed that because their first bump-up in income made them happier, having still more money would further expand their happiness. It's like having two beers and feeling good and then downing ten beers, believing that this will make you feel five times happier. It doesn't work that way.[37] The stuff that big money buys—fancy cars, trophy houses, exotic vacations—often leaves people feeling entrapped; rather than *having* stuff, their stuff ends up *having* them.

Sadly, one thing that *has* increased in the United States in tandem with rising consumption is the incidence of depression, now ten times higher than in the 1950s.[38] Psychologist David Myers refers to this pattern of rising wealth and diminished well-being as "The American Paradox." Specifically, Myers observes that at the outset of the twenty-first century, Americans found themselves "with big houses and broken homes, high incomes and low morale, secured rights and diminished civility. We were excelling at making a living but too often failing at making a life. We celebrated our prosperity but yearned for purpose. We cherished our freedoms but longed for connection. In an age of plenty, we were feeling spiritual hunger. These facts of life lead us to a startling conclusion: Our becoming better off materially has not made us better off psychologically."[39]

Summing up, economism has many troubling effects on us—both as individuals and as a society—but the main message here is that economism is simply a story and we suffer its negative effects only insofar as we accept its version of reality. Laying bare the existence of this story, and its effects, is a necessary prerequisite for creating and living within a new more life-affirming and ennobling story.

STEPPING BACK TO SEE THE BIG PICTURE

> You see everything is about belief, whatever we believe rules our existence, rules our life.
>
> —Don Miguel Ruiz[40]

The general ideological thrust of Western culture has continued largely unchanged for the last thousand years. In the Middle Ages, prior to the Renaissance, Western life was organized primarily around religion. The basis for governance was formed around religion; works of art were frequently created to glorify God; and armies were assembled and wars waged mostly to further religious agendas. The emergence of Renaissance humanism (fifteenth century) was a revolutionary reaction against the constraints imposed by Church and state power in the medieval world. Humanism posited that man, by virtue of his rational mind, had the capacity to understand the workings of nature. By the sixteenth and seventeenth centuries, scientists such as Copernicus, Galileo, and Newton were demonstrating that it was possible to ascertain reliable knowledge of the workings of the physical world. Gradually, the medieval story—about salvation through divine grace—was replaced by the doctrine of material progress. In many quarters, man, not God, became the central focus of life; and reason triumphed over dogma.

The Enlightenment (eighteenth century) represented the consolidation and extension of the humanist-scientific orientation to all spheres of human endeavor. The "medieval Christian world, wherein all creation was infused with God's presence and direction, was replaced by a sense of a clockwork-like universe that had been set in motion by God in the beginning but otherwise operated autonomously, according to Newtonian laws."[41] By the end of the eighteenth century, modern man had emerged as the detached manipulator of the world, with a strong secular-rationalist orientation centered on the pursuit of self-interest. Once scientists had made it abundantly clear that Earth was not at the center of the universe, Enlightenment philosophers elevated the individual to that exalted position.

But the expansion of secularism, since the 1500s, isn't so much a new story as it is a new spin on Western culture's old story. If anything, modern secularism has strengthened the tendency for us to see ourselves as the culmination of evolution and to believe that it is our destiny to control and manipulate nature.

Economism, the present permutation of Western culture's story, is based on the notion that Earth is primarily a lump of resources waiting to be transformed into products for our use. This worldview presents a depressing characterization of the human enterprise, suggesting that work, money, and material goods are the central reason for—and significance of—our lives.

We can choose another story. In fact, every day each of us creates stories as we try to make sense of the world. For example, suppose you are seated in a restaurant and a friend comes in and fails to acknowledge you. How will you interpret this? That your friend didn't want to disturb you? That she is mad at you for some reason? That she forgot to put in her contacts and didn't recognize you? Each interpretation is a story even though—when you are caught in its grips—it feels like the absolute truth. To underscore this point, consider this vignette from Byron Katie's book, *Loving What Is*:

> Once, as I walked into the ladies' room at a restaurant near my home, a woman came out of the single stall. We smiled at each other and as I closed the door she began to sing and wash her hands. What a lovely voice! I thought. Then, as I heard her leave, I noticed that the toilet seat was dripping wet. How could anyone be so rude? I thought. And how did she manage to pee all over the seat? Was she standing on it? Then it came to me that "she" was a man—a transvestite, singing falsetto in the women's restroom. It crossed my mind to go after her (him) and let him know what a mess he'd made. As I cleaned the toilet seat, I thought about everything I'd say to him. Then I flushed the toilet. The water shot up out of the bowl and flooded the seat. And I just stood there laughing. In this case, the natural course of events was kind enough to expose my story before it went any further.[42]

It is easy to get caught up in stories that cause suffering, as was the case for Byron Katie. It's the same with the story of economism. Caught up in this story, suffering comes in the form of separation from each other, from planet Earth, and from the abundance and well-being and happiness that is our birthright. But none of this is necessary. Just as Byron Katie awakened to her enfeebling story when she flushed the toilet and saw the water shoot up, we awaken to the crippling effects of economism as we face its limitations and summon the courage to imagine a different story.

I sometimes offer my students an embodied experience of what it feels like to shift from an old deadening cultural story to a new enlivening one by placing a rope on the floor in the front my classroom and then inviting five volunteers (the "A" group) to stand on one side of the rope facing a second group of volunteers (the "B" group) on the opposite side of the rope. Once everyone is ready, I ask the "A" volunteers to greet their partners with a feisty "Yo Mama!" I even have them practice their "Yo Mama" until they can say it with attitude. Then, I invite the "B" group to respond with, "Don't you *Yo Mama* me!" Next, I tell them to reach out and grab their partner's right hand. Then, I inform them that their task is to "do anything in their power to move their partner across the rope." They will have only five seconds to do this. The "As" get ready by repeating "Yo Mama!" and the "Bs" respond with, "Don't you *Yo Mama* me!" Then, I shout "Go!" and while each participant struggles to get their partner across the rope, I count down: 5 . . . 4 . . 3 . . . 2 . . . 1. Usually, one or two people in each pairing manage to drag their partner to their side of the rope. While they stand panting, I ask who feels like he or she won or lost. And then, after the winners and losers acknowledge as much, I invite them to do the task a second time, but this time to approach it so that they both win. As they line up, I slowly repeat the instruction: "Your assignment is to do anything in your power to move your partner across the rope," and, then, I shout "Go!" This time most pairs merrily glide past each other, voluntarily exchanging positions. There are no overt winners or losers, just people who have completed a task in a mutually satisfying manner.

Sometimes, a student will complain that I set them up to view this as a competition by goading them to say "Yo Mama" and giving them only five seconds to complete the task. I agree and explain that I was purposely re-creating what our culture often does to them—it encourages them to see life (including their time in school) as a competition or race to the finish, with winners and losers. Finally, I invite them to consider what it would take to see themselves and each other with new eyes—each as unique manifestations of life, each a partner rather than a foe, each a fellow human to extend kindness to rather than to be threatened by. In this way I am inviting students to try on a new story—one grounded in cooperation rather than competition; one centered on the generative question, What can we do for each other? rather than the soulless refrain of economism, What can I get for myself?

WRAP-UP: EXPANDING ECOLOGICAL CONSCIOUSNESS

Tell me the story of the river and the valley and the streams and woodlands and wetlands, of shellfish and finfish. Tell me a story. A story of where we are and how we got here and the characters and roles that we play. Tell me a story, a story that will be my story as well as the story of everyone and everything about me . . . a story that brings us together under the arc of the great blue sky in the day and the starry heavens at night.

—Thomas Berry[43]

In this chapter we have entertained the possibility that Western culture's story—a story founded, to a significant degree, on conquest, control, dualism, hyper-individualism, scarcity, acquisition, separation, and more—is at the root of today's socio-ecological malaise. In its modern guise of economism, this story proclaims: Humans are essentially economic beings and the subjugation of nature is necessary for progress. These are the truths that we are all taught—American, Russian, Chinese, and Colombian—but they are, I fear, only half-truths, unable to ensure the full flourishing of life.

But the solution is not to shun modernity. This would be folly. As philosopher Ken Wilber reminds us:

> The rise of modernity . . . served many useful and extraordinary purposes. We might mention: the rise of democracy; the banishing of slavery . . . the widespread emergence of empirical sciences, including the systems sciences and ecological sciences; an increase in average life span of almost three decades . . . the move from ethnocentric to world-centric morality; and, in general, the undoing of dominator social hierarchies in numerous significant ways. These are extraordinary accomplishments, and the anti-modernist critics who do nothing but vocally condemn modernity, while gladly basking in its many benefits, are hypocritical in the extreme.[44]

According to Wilbur, the time has come to take the good that modernity offers and leave behind its harmful effects, attitudes, and behaviors.[45] Indeed, another, more life-affirming story is possible. For example, we could choose a story like:

> We humans are born to connect, to learn, and to create in ways that respect and honor Earth; and it is in our very nature to seek to create a world of cooperation, sharing, justice, beauty, love, abundance and peace.

If we were to live within a story like this, then economism's soulless *trinity* of Market, Money, and Technology would be replaced by a more life-affirming *trinity* consisting of Community, the genuine source of security; Love, the force that ensures connection and genuine well-being; and Spirit, the generative power that serves as the ground of existence. Insofar as we, as a people, can change our story, we can change our destiny.

APPLICATIONS AND PRACTICES: DISCOVERING THE POWER OF STORY

> Who would you be without your story? You never know until you inquire. There is no story that is you or that leads to you. Every story leads

away from you. Turn it around; undo it. You are what exists before all stories. You are what remains when the story is understood.

—Byron Katie[46]

Our most important story—the one that influences everything else in our lives—is the story we have about ourselves. Who are we? How did we get here? What is our purpose? What is our relationship to Earth and cosmos? Our answers to these, and related questions, comprise our personal belief system or worldview.

Because we live so completely within our particular worldview, it is hard to actually see it. The movie *The Truman Show* is instructive in this regard. In this film Jim Carrey plays the part of a man (Truman) whose entire life is a television show that is broadcast around the country. The remarkable thing is that Truman does not know this—he is just living his life, or so he thinks. The director of the television show, in effect, orchestrates Truman's life, determining when the sun will shine or clouds will appear and what occurs in Truman's life on any given day. During one episode in the movie a reporter asks the director why it is that Truman is not able to figure out that his whole life is just a TV show. The director responds candidly, "We all accept reality as it is presented to us."[47] What about you? Are you consciously living your life or has your life become a deeply habituated thought pattern?

Discovering Your Worldview

Whether consciously articulated or not, our beliefs extend into every pocket of our existence, guiding our actions and decisions, from life till death. For example, if you have been conditioned to believe that birth should occur in a hospital, children should go to school, adults should get married and have families, old people should retire, and dead people should be buried underground, then you will follow this life script. Yes, our beliefs (how we see!) determine our destiny. The point, of course, is that without conscious awareness of our beliefs, we aren't nearly as free as we imagine ourselves to be.

It is possible to begin to identify the beliefs that comprise your worldview by responding to open sentences such as the following:

1. _____ created the world.
2. The universe is _____.
3. Earth is here to _____.
4. Earth's plants and animals are here to _____.

5. Human beings' purpose in being here on Earth is to _____.
6. The good and bad things that happen on Earth are the consequence of _____.

Your responses to these six open sentences are like puzzle pieces that comprise parts of your worldview—that is, the story that influences, at least in part, the purpose, direction, and meaning of your life.

And if you are want to explore how new ways of seeing might open you to new possibilities, you could note down answers to the above six open sentences a second time, but this time responding in such a way that you create an entirely different story. So, for example, if, in the first round, you wrote, "*God created the world*," in your second round you could conjure another possible answer to this question. The idea—and this is the important part—is to make your second set of responses more mind-expanding and exciting than your first set. It is not necessary that you believe your second responses, just that you exercise your innate creativity—your fecund imagination—to come up with new possibilities that—though they may be a bit far-fetched—create a measure of excitement and intrigue within you and, in so doing, have the potential to shift your consciousness toward a new story.

If you want to go deeper still, perhaps give thought to these four questions: (1) How does your second *worldview/story* leave you feeling? (2) What do you believe would happen if you began to act as if this new story were true? (3) What would you lose? (4) What might you gain?[48]

Discovering Stories in Your Daily Life

The Inuit people have a saying that the Great Spirit must have loved stories, for why else would he have created so many people? Each of our lives is, indeed, a story, with each day a chapter.

You can, if you choose, adopt simple practices to awaken to the stories in your life. For example, you could see the morning newspaper as a storybook; or regard all your daily encounters and activities as having a beginning, middle, and end; or see each meal as a story involving food preparation, eating, and cleaning up; or listen to the eleven o'clock news as a goodnight story; and so forth. To formalize this practice, you might even consider doing an "evening review" of the events of your day. In other words, run the movie of your day through your mind. As you do this, pay particular attention to the activities that you engaged in, the interactions you had with others, and the thoughts and conversations surrounding any

STORY HARVESTING

I was waiting to board an airplane. The line was long and moving slowly. As I stood in a state of impatience, I watched a stewardess guide an elderly couple down the aisle. They were both large people, and it appeared as if the man had suffered a stroke because his speech was slurred and his posture bent. A few minutes later, when I approached my seat at the back of the plane, I noticed that the same couple was seated in my row. I confess, now, with some embarrassment, to feeling irritated at the thought that I would have to squeeze in next to them, but, as it turned out, my seat was on the other side of the aisle.

Later, about halfway through the flight, a stewardess came down the aisle holding up a $20 bill, asking for change. It occurred to me that I might have change, but I decided that it would be too big of a hassle to undo my seatbelt and dig my wallet out of my back pocket to search. Just after I reached this decision, I heard the man with the garbled speech say that he thought he had change. It was the man I had earlier wished to distance myself from. For several minutes he wrestled with his seatbelt and then his wallet. Finally, he produced change for the twenty, and the stewardess thanked him.

It wasn't until later, when I paused to review the events of my day, that I was able to fully consider the significance of this event. I recalled my chagrin when I thought that I would have to squeeze in with the oversized couple; and then I recalled my deep humiliation because it was that bent-over man who had helped the stewardess when it would have been so much easier for me to have done so. That man was my teacher that day, and I was grateful for his lesson in courtesy and kindness.

decisions you made during your day. Think of this evening review as a way of harvesting the stories of your life.

Author Greg Levoy recommends pausing from time to time to embrace your entire life story as a grand myth—an epic story of your destiny and travails. Why do this?

> [Myths] get at the heart of human behavior, at profound truths, universal truths, ageless patterns. They are . . . stories of transformation: from chaos to form, sleep to awakening, woundedness to wholeness, folly to wisdom, from being lost to finding our way.[49]

In the end, each of us has a story, uniquely our own to enact. But it is easy to forget this and become subdued by our culture's insistent pleas to:

- Be careful!
- Look out for yourself!
- Avoid unnecessary risks!
- Control your passions!
- Obey the rules!
- Be polite!
- Conform!

These are the messages generated by the limiting story of economism. To the extent that we acquiesce—sacrificing our souls for a saccharine security—we risk having the light of spirit and consciousness grow dim within us.

QUESTIONS FOR REFLECTION

- What are the stories that you were told when you were growing up, and how have these stories shaped your understanding of the world?
- What are the messages and beliefs underlying the stories you receive from the news media? How do these messages leave you feeling?
- In what ways is economism similar and dissimilar to other "isms" such as nationalism, sexism, patriotism, and racism?
- What is your experience of separation in your daily life? And related to this, when, where, and to whom do you feel connected and what provides you with that feeling?
- If economism is Western culture's present story, what is the story that is waiting to be born?
- Do you live in a world of scarcity or abundance? How do you know?

8

BIRTHING A NEW STORY

The Great Turning

It was an old story that was no longer true. . . . Truth can go out of stories, you know. What was true becomes meaningless, even a lie, because the truth has gone into another story.

—Ursula K. Le Guin[1]

In her book *In the Tiger's Mouth*, Katrina Shields tells a story about a young Irish woman with no political background who was working as a clerk in a Dublin supermarket. The woman had read about the apartheid regime in South Africa—whereby the white minority was subjecting the black majority to political, legal, and economic discrimination; and she had heard about the call from South African blacks to boycott South African produce. When she discovered that her store carried South African fruit she was dismayed. So it was that she simply decided to stop selling it. She dutifully rang up all the other groceries, she refused to sell the fruit from South Africa. This woman was saying "no" to the old story of apartheid, with its attendant separation, thereby aligning her personal values with her workplace behaviors.

What happened next? Well, not surprisingly, customers complained to the manager and, in short order, the young woman was fired. But by the end of the day, all the other clerks in the store also were refusing to sell the South African fruit. And soon workers in other supermarkets around Dublin were coming out in support of the women who refused to sell the "racist" fruit. There were threats of mass shutdowns, and the media jumped on the story. In the end, the supermarket stopped selling South African fruit and the woman got her job back.[2]

Like the woman from Dublin, we each have opportunities, if we choose, to say "no" to stories that are in violation of our values, our vision, our integrity. When I turned fifty-five, my partner, Dana (forty at the time), expressed her deep desire to become a mother. At first, I rejected her suggestion categorically, believing that I was too old, that people (including my own grown children) would judge me as foolhardy and irresponsible, that life should be lived in a certain order, and that it was a mistake to repeat life phases. But then I began to investigate my story, asking questions like: Who says I am too old? Who says it is foolhardy? As I did this, I was able to create a new story—one not contaminated with the fear of how I might be judged by others, but, instead, stabilized in trust and love.

It's the same today, I believe, for humanity as a whole. We are outgrowing our old story. It served us for a time but no longer. We are now being challenged, just as people were in the time of Galileo and Copernicus, to adopt an entirely new worldview. Pain and bewilderment (as well as excitement) surface anytime a person or a people experience the dissolution of their old worldview—the story that heretofore directed their actions; and it is no different in our time. No single one of us knows with any certainty what lies ahead but many concur that we live in a fecund time pregnant with chaos, surprise, and grand possibilities. In this final chapter, I invite you to join minds and hearts with hundreds of millions of others around the world in the birthing of a new story. How? Simply by paying attention and looking and listening for the signs of the Great Turning—i.e., humankind's *turning away* from crippling separation and *turning toward* kinship, community, and deeper purpose. Indeed, this Great Turning is already well under way.

FOUNDATION 8.1. WE ARE MADE FOR RELATIONSHIP

Pick any environmental or social problem: climate change, species extinction, environmentally linked cancers, ocean acidification, war, aquifer depletion, genocide, deforestation, loss of soil capital, coral reef decline—and, in every case, it is separation—the severing of relationships—that is at the root of the problem. But take heart: The new story, emerging all around us, is the antidote to separation insofar as it challenges us to see Earth, each other, and ourselves with new eyes—the eyes of interdependence.

Seeing Earth with New Eyes

In the last chapter I posited that, from the perspective of economism, Earth is a kind of warehouse that humans extract things from whenever they want. In this story, Earth is an object that is separate from us.

Not surprisingly, our separation from Earth is often revealed in our manner of speaking and writing. For example, in English we routinely use the article "the" before the word "earth" (spelled with a lower case "e") saying or writing things like "The earth is the third planet out from the sun." This seems like correct speech and yet, when referring to Mars, we would never say "*the* mars." By writing or speaking "the earth" we convey, in effect, that Earth is something that is separate from us—that is, we are here and "the earth" is there. As Unitarian minister Michael Dowd points out, "To refer to the literal ground of our being, the source and substance of our life, as 'the earth' is to objectify it. Such objectification encourages us to continue seeing Earth merely as a resource for human consumption."[3] In contrast, when we use our planet's proper name, "Earth," we honor its integrity as a creative, self-organizing, living presence that animates life in all its forms.

To further pursue how language can affect our perceptions of Earth, consider the elementary question, Do you live *on* planet Earth or *in* planet Earth? Which is it? Most people respond that they live *on* Earth. After all, saying that "I live *in* Earth" would sound a bit weird. But the truth is that all of us actually live *in* Earth! If you doubt this, go outside and lie on your back and then look up at the sky. In particular, observe the clouds, with the full knowledge that Earth is spinning in an easterly direction at more than a thousand miles per hour. Given this fact, it stands to reason that those clouds up there should be zooming across the sky toward the west at a thousand-plus miles per hour. But they are not. Why? Because all that stuff up there that we blithely refer to as "sky" or "atmosphere" is part of planet Earth. Yes, that's planet Earth up there too! If you still doubt this, hold your breath! Just as a fish is utterly dependent on water for life, our support medium is air.[4] In fact, when our aquatic ancestors moved from the sea onto dry land hundreds of millions of years ago, they were, in effect, transferring their allegiance from a watery ocean below to a gaseous ocean above.[5] Every day, every second, each of us is intimately entangled with Earth's atmosphere—the surfaces that comprise our skin, lungs, internal organs, tissues, and muscles are imbued with atmosphere and inseparable from it.

As a language-based way of acknowledging that the air that enfolds us is part and parcel of Earth, David Abram has suggested that we change the spelling of Earth to E*air*th. We might even think of the upright "i" in the middle of the word E*air*th as symbolizing the human—the "I" standing upright, sustained by air above and soil below.[6]

Taking all this in, it is easy to see that the expression "Mother Earth" is not cute or fanciful but rather an accurate description for what Earth is to us—the womb of our very lives and the lives of our ancestors and of our children and their children to come. In this vein, Richard Nelson writes:

> There is nothing of me that is not earth, no split instant of separateness, no particle that disunites me from the surroundings. I am no less than . . . earth itself. The rivers run through my veins, the winds blow in and out with my breath, the soil makes my flesh, the sun's heat smolders inside me . . . the life of . . . earth is my own life. My eyes are . . . earth gazing at itself.[7]

Nelson's imagery—Earth's winds in our breath; Earth's waters in our veins—is enchanting. Enchanting, that is, until, with a gasp, we realize that Earth's winds and soil and water—and by extension our very breath, body, and blood—have become laced with myriad toxins because of human hubris, ignorance, and alienation (see chapter 5).

In light of our predicament, campaigns to save the environment, even when successful, may end up being no more than holding actions, insofar as they address only symptoms and not root causes. The really big task ahead, it seems, is not to fix the environment but, rather, to fix our ways of seeing and, by extension, our ways of thinking—in other words, our story!

Fixing our ways of seeing means coming to understand that Earth *is* our larger body, that when we walk along a sidewalk or forest path, we *are*, quite literally, Earth walking. You are, I am, each of us is a manifestation of Earth. This realization, when we really get it—I mean get it in our bellies and bones—can't help but transform how we see ourselves in relationship to Earth and its myriad creatures.

By way of illustration, Joanne Lauck's story (next page) suggests that in extending kind regard to insects we might, actually, become more peaceful and loving toward each other. It appears that the Dalai Lama agrees because when he was asked what he thought was the most important thing to teach our children, he responded, "Teach them to love the insects!" This is wise counsel, especially insofar as many children these days have been conditioned to fear insects—seeing them as annoying and scary. But just think: If all of us—adults and children alike—could learn to appreciate and respect insects in their myriad and mysterious forms, we just might be able to do the same or better when it comes to respecting and loving each other!

EXTENDING RESPECT TO ALL OF EARTH'S CREATURES

Joanne Lauck, in her book, *The Voice of the Infinite in the Small*, tells a story about Joanna Macy, a teacher-activist who traveled to a village in Tibet to help the people there start a craft cooperative. One day during her visit Macy met with a group of Buddhist monks. During the meeting a fly fell into Macy's tea. Though this was not a matter of concern for Macy, the young monk who was sitting beside her leaned over and asked if she was OK. Macy responded that she was fine. After all, it was only a fly. A moment later the monk again expressed his concern and she reassured him that she really was OK and he need not concern himself with her. Finally, the young man, overcome with concern, fished the fly out of Macy's tea and disappeared. Out of sight, he gently placed the tea-soaked fly on a bush until it had revived. When he returned to the meeting a few minutes later, he was beaming as he assured Macy that the little fly was going to be OK. For Macy, this was a poignant moment as she came to see that the young man's compassion extended not just to her but also to the fly, the lowly fly!

In interpreting this story, Lauck writes: "If something sputters inside us in indignation at the thought of helping a fly, it may only be self-importance arising out of a narrow band of awareness. Our sense of self expands when we extend our compassion to insects. It is an appropriate response to our interdependence. The question is not how to connect with a fly—we are already connected. The question is how to translate that connection into appropriate behavior. Helping one in need is always a good place to start."[8]

Seeing the Human *Other* with New Eyes

A quick glance at modern history reveals that, though humans are capable of kindness and love, we often fail to exhibit these qualities. Now, at the outset of the third millennium, it is more evident than ever that oppression and warfare are not effective means of achieving enduring peace or wellbeing. It's taken a long time but we are finally coming to see that "The key to changing our relationship to the *Other* is to expand our notion of *Self* to include everyone [we] see."[9]

Recent research reveals that human beings are actually more attuned to each other than previously imagined. For example, Daniel Goleman in his book, *Social Intelligence*, reports that researchers in Sweden have discovered that simply seeing a picture of a smiling face is enough to subtly activate the smile muscles of our face. More generally, whenever we see a photograph of someone displaying a strong emotion, whether it be happiness, grief, rage, or anything else, the muscles of our face automatically begin to mirror the emotion we are observing and in turn these subtle shifts in our facial muscles actually cause shifts in our feeling state.[10]

This response is neurological. For example, if I were to prick your finger with a pin, neurons would light up in a certain part of your brain. This is to be expected, but what has astonished scientists is that your brain would also light up in the same place and way if you watched as I pricked someone else's finger. Indeed, it is now known that the human brain is wired with a class of cells dubbed "mirror neurons." These neurons fire in response to the emotions that the people around us are displaying, in effect, grafting their emotional life onto ours.

Goleman puts it this way: "Mirror neurons ensure that the moment someone sees an emotion expressed on your face, they will at once sense that same feeling within themselves. And so our emotions are experienced not merely by ourselves in isolation but also by those around us—both covertly and openly."[11] We are, it seems, wired for empathy, or as Marc Barasch puts it, "I feel you in me."[12] In sum, mirror neurons "suggest that an archaic kind of sociality, one which does not distinguish between self and other, is woven into the primate brain."[13] It is as if the Golden Rule is, in some measure, inscribed into our very biology.

It also appears that we are born predisposed to cooperate. For example, Michael Tomasello in his book, *Why We Cooperate*, reports that when small children, less than two years old, see that an adult (even one they have just met) is unable to open a door because his hands are full, they will rush to his assistance. In a similar vein, if an adult pretends to lose an object, a one-year-old will point him to it; or if the adult drops an object in front of a young child, the child will instinctively pick it up and give it to the adult without any request to do so. Tomasello makes the case that such cooperative/empathic behaviors are not taught or performed for a reward; rather, young children are cooperative by nature.[14] However, by the time kids reach the age of three their natural inborn predisposition for cooperation may be thwarted if they live in a culture (like that in the United States) that often tends to discourage collaboration in favor of individualism.

It doesn't have to be this way. In some indigenous cultures a sense of belonging and interconnectedness shapes all human relations. For example, Martin Prechtel, a Native American who lived for many years among the Tzutujil people in Guatemala, relates that in his village if a family had a sick child, the family as a whole understood itself to be sick. Likewise, if several people in the village were sick, the village, as a whole, understood itself to be sick. In fact, this same collective consciousness even extended to the land. For the Tzutujil it was impossible to say, "I am healthy but the village forest is dying." For them, to imagine that they could be healthy when their family, their village, or the surrounding forest was not, would be as absurd as you or me saying, "I've got a fatal liver disease, but that's just my liver—I'm really fine!" Just as your sense of self includes your liver, theirs includes their family, village, and surrounding land.[15]

Embracing, rather than rejecting, the human *Other* represents a dramatic departure from the predominant story line of human history—namely: to project blame onto the *Other*, always imagining that someone else is the problem. Indeed, much of human history is the story of who killed whom—a tale of vengeance. This carnage has often been what anthropologists call "sacred violence"—violence justified in the service of a holy cause. But, in

---

DISSOLVING THE BOUNDARIES
BETWEEN SELF AND *OTHER*

You don't have to be an indigenous Tzutujil from Guatemala to experience the *Other* as a kindred spirit. For example, St. Louis native Angie O'Gorman was awakened late one night by an intruder standing over her bed. No one else was in the house. Somehow, amid her fear, she realized that there was a connection between herself and this *Other*—either they would both be damaged by what ensued, or they would both emerge safely, with their integrity intact. This insight allowed O'Gorman to act with empathy for both herself and the intruder.

She spoke calmly into the darkness, asking him what time it was.

"Two-thirty," he replied.

She expressed concern that his watch might be broken because the clock on her nightstand read 2:45. Then, after a pause, she asked how he had entered her home. He said he broke a window. She said this was a problem for her because she didn't have the money to fix it. He confessed that he was also having money problems.

When she felt it was safe, O'Gorman told him, respectfully, that he would have to leave. He said he didn't want to . . . had no place to go. Lacking the force to make him leave and seeing a person without a home, O'Gorman said that she would give him a set of sheets but he would have to make his own bed downstairs. The man went downstairs, and O'Gorman sat up in bed for the rest of the night. The next morning they had breakfast together. Then the man left. By cultivating empathy for this *Other*, O'Gorman entered a new plane—a place where no one wins unless everyone wins, because if someone loses, everyone loses.[16]

our time, such redemptive violence, finally, is being seen for what it is—a misguided attempt to avoid our own pain and brokenness. As theologian Richard Rohr points out, "If you don't transform your pain, you will always transmit it. . . . You will always find someone who is worthy of your hatred . . . and then, of course, you become the evil you despise."[17]

Ultimately, we are at a moment in history when we are striving to make a shift from a consciousness based on exclusivity to one based on inclusivity. This monumental shift necessitates that we *see with new eyes*. Inclusivity is grounded in relationship, whereas exclusivity stems from separation. A consciousness rooted in inclusivity generates trust; one moored in exclusivity foments fear—especially fear of the *Other*. When our goal is exclusivity, we silence those with whom we disagree; but when inclusivity becomes our goal, we seek to create a world that works for all.[18]

Seeing Ourselves with New Eyes

In the end, it all comes down to how we see ourselves. If we believe that we are fundamentally separate from each other and Earth, we will live in ways that engender precisely this separation.

Certainly, our culture—grounded as it is in performance and competition and individualism—leads us to see ourselves as apart from each other and Earth. For example, we are encouraged to think of ourselves in terms of our name, our beliefs, our possessions, and so forth. But if you stop to think about it, you will discover that you are really not your possessions, not your occupation, not your body or your physical appearance, not your emotions, not your thoughts. "Wait a minute," you might be thinking, how could I not be these things? For example, how could it be that I am not my thoughts? Well, for one thing, if you are like most people alive today (including me!) most of your thoughts are not really yours so much as opinions and thought patterns that you have absorbed from the world around you—from your parents, teachers, the media, religion. That we so readily identify with our thoughts and mistake them for who we are can lead to a tragic misidentification of self.

Indeed, if you pay attention, you will realize that your thoughts are changing from minute to minute. There is nothing permanent about them. They come and they go. The same is true for your body. You are not your body. Your body, too, comes and goes—millions of cells dying each minute and new cells forming. If you had taken an inventory of the trillion-plus cells in your body a decade ago and then were to re-census your body cells today, you wouldn't find one of those old cells still alive! The cells of your body don't persist, but the sense of *selfness* that you have known since you were small stays with you.

What about other aspects of your identity? How about your name? Is your name who you are? Clearly not, because you can change your name and you will still be you. And what about your religion? If you drop your old religion and get a new one, will you have a new "I"? No, it will be the same "I," simply with some new attitudes, insights, and beliefs tacked on. So go ahead and change your clothes, your car, your college, your profession, your body, your GPA—change all those things and it will not change who you most essentially are. Indeed, it is curious that most of the things that we say that begin with *I, me*, or *mine* are just surface attributes—having little to do with the essence of who or what we are.

The upshot of this sort of inquiry into personal identity is that most of us have a very superficial understanding of who we are. It is as if we have chosen to ignore the basic dictum, "Know thyself," passed down to us from the ancient Greeks. The ancient ones knew what we moderns seem to have forgotten—namely: We can't truly know ourselves until we break free from our socially conditioned ways of seeing ourselves.[19]

The disorienting consequences of misperceiving ourselves are illustrated in a story told by Anthony DeMello about a man who found an egg of a wild bird and then placed it in the nest of a barnyard hen. Eventually the egg hatched and out came a baby eagle. The eaglet was reared with a brood of chickens, and followed their example, scratching the ground for worms and

SEEING AND BEING SEEN

With the intention of promoting self-awareness, I sometimes invite students to form pairs and then to simply gaze into each other's eyes without speaking. Some find this too challenging to even attempt; others begin but after a few seconds break down in nervous laughter. A very few are able to actually calmly gaze into their partner's eyes demonstrating that it is possible. Full-grown adults struggle equally with this task, prompting the question, Why is it so difficult for so many of us to simply look, with calmness, into another's eyes? I mean, there we all are—fellow humans—and yet we often have difficulty beholding each other without becoming agitated. Doesn't that seem a bit strange? Are we afraid of what we might see? Or, perhaps, afraid of being seen? Might our struggle with this task say something important about how our culture separates us from our very selves by conditioning us to see ourselves as never OK just as we are, never good enough, never quite worthy of appreciation and acceptance?

insects, clucking and cackling, and thrashing its wings in a feeble attempt to fly. Then, one day, when the eagle was several years old, he looked up and was awestruck by a magnificent bird flying high in the sky. He asked a nearby chicken, "Who is that?" The chicken responded, "That's the eagle, the king of birds; he is different from us; he belongs to the sky. We, on the other hand, are chickens; we belong to the ground." So it was that the barnyard eagle continued to live as a chicken until the day he died. Why? Because that is what he believed he was.

Might it be the same for us—that is, that our culture has conditioned us to see ourselves as small, limited, and flawed beings meant to spend our lives as drones with our heads to the grindstone when, in fact, we are creatures infused with goodness, compassion, and creative genius—capable of things beyond our wildest imaginings?[20]

FOUNDATION 8.2: A NEW STORY: COMMUNITY AS LIFE'S FOUNDATION!

The story of the United States—its creation and history—though pocked with darkness in some places, has had an overall ascendant quality. As Yale University professor Gus Speth narrates, it is a story of "a people who set out on a journey—a journey through time—to build a better world for themselves and their children. High-minded and full of hope as they began, they accomplished much in their quest. But they became so enamored of their successes, indeed captured by them, that they failed to see the signs that pointed in new directions, and they became lost."[21] This seems like an apt description. We are a people in search of a meaningful narrative—one with the power to rouse and awaken us, to point us toward our true nature, to bring meaning and deep purpose to our lives.

Fortunately, a new story—a new worldview—is emerging, albeit slowly.[22] This new story is inviting us to see ourselves not as fighters on the battlefield of life but as lovers in an unfolding universe; not as cogs in a machine, but as a communion of subjects intimately related to each other; not as selfish and brutish sinners, but as fundamentally kind and compassionate beings of light; not as drones meant to spend our lives with our noses to the grindstone, but as beings on a journey to ever-deeper awareness and broader consciousness; not as greedy characters living in the pursuit of more belongings but as relational beings, dwelling in a world of abundance.

It is heartening to realize that we are not nearly so stuck as it might at first appear. Cultures do change. Things that now seem normal—such as

---

**FROM THE PAIN OF SEPARATION
TO THE BLISS OF RELATIONSHIP**

On a recent forest field trip with the students in my ecology class I decided to bring along a snack consisting of nuts and raisins, some grapes, and a few cookies. After setting the treats out on a picnic table, I told the students to find a straight stick about the length of their arm. When they returned with their sticks I gave them two thick rubber bands and instructed them to secure the stick to their right arm so that they could no longer bend it. Then, I invited them to partake in the food set out on the table, with one proviso—that they hold their left arm behind their back and use only their right arm (the one that they could not bend) for eating. They laughed, knowing this was a game of some sort and, then, they set about eating their snacks. Unable to bend their eating arm, some tried to throw the food up in the air and catch it in their mouths; others sheepishly put their mouths to the table and vacuumed up snippets of food. None of this was very easy or pleasant. Noting that they were becoming more frustrated by the moment, I suggested that there was a way to transform their frustration into happiness, but it would require that they take a risk. That's all it took. Soon they were all using their straight arms to feed each other. This seemed like a "heaven-on-earth" experience insofar as everyone was receiving twice over—once when they were fed by a classmate and a second time when they were the ones offering food. The point, once again, is that how we perceive reality—as relation based or separation based—can determine our behavior and our happiness.

---

hyper-individualism and economism—can suddenly come to be seen as untenable and brought to an end with surprising quickness. These seismic changes occur as people free themselves from limiting beliefs and in the process cultivate a new worldview.

The Purpose of an Economy Is to Serve the Community

It seems that we have forgotten what our ancestors knew very well—namely, that the purpose of an economy is to serve people. That's right, an economy, in its simplest formulation, is a network of relationships created to serve people so that they might live well, securely, and happily. With this in mind, economist David Korten has called for a shift to "a life-serving economy" where the goal is to generate a good living for everybody rather than vast wealth for a few.[23] In such an economy, our soils would be respected (rather than abused), our waters would be clean (rather than polluted), our climate would be steady (rather than destabilizing), our families would lead meaningful and joyful lives (rather than lives marked by stress and strife), and our spirits would be animated (rather than subdued).[24] The centerpiece of a life-serving economy is no different than the centerpiece of life itself—community!

Genuine communities—where people know and look out for each other—possess a unique form of wealth known as "social capital." In such communities, neighbors trust each other; share talents and resources with each other; engage in problem solving together; and, not surprisingly, are there for each other in times of hardship. Such communities—rich in social capital—are able to transcend, to a significant degree, the limitations imposed by limited access to financial capital. In this sense, social capital acts as a genuine insurance policy—something people can truly count on in the event that their retirement assets disappear in a financial meltdown.

Research reveals that people living in communities rich in social capital tend to be healthier and happier—that is, better able to cope with trauma,

SOCIAL CAPITAL AND HUMAN LONGEVITY

We often think that the key to human health and longevity is a healthy diet and exercise. In this view, health becomes a personal project—something we can attend to privately. However, a large body of research reveals that we may get a better health result by simply joining a club, helping others, spending time with friends, and/or participating in a spiritual community. Public health practitioners sometimes refer to the health benefits resulting from strong social bonds as the "Roseto Effect," a reference to a neighborhood of poor Italian immigrants living in Roseto, Pennsylvania, in the 1950s. Doctors, at the time, were surprised that the folks in Roseto were more long-lived than people in surrounding communities, even though folks in Roseto displayed the same health-risk factors (e.g., frequent smoking, diets high in animal fat, overweight, lack of regular exercise). The only significant difference they could find had to do with community bonds. The Italians in Roseto were distinct insofar as they cared for and helped one another, manifesting much more community spirit and interdependence than in surrounding places. In the Roseto of the 1950s what mattered most was relationships.

Today, the children of those early Italian immigrants have slowly become indistinguishable, in many ways, from typical Americans—that is, they have become adherents of economism—more concerned with personal goals and the acquisition of things than with maintaining strong networks of interdependence. The result: Community bonds in Roseto have declined, and so, apparently, has the Roseto Effect.[25]

resist illness, and less likely to suffer from depression.[26] So it is that sociologist Eric Klinenberg found that the people who died in the 1995 Chicago heat wave were similar in one respect: they were socially isolated. In a related vein, those people who had a solid support network prior to the Hurricane Katrina disaster tended to fare better than those lacking such social capital.[27] Social connections, it appears, really are insurance that we can count on.

Even absent an immediate crisis, simply being part of a genuine community contributes to human well-being. For example, when researchers at Carnegie Mellon University attempted to infect people with a cold virus by spraying the virus directly into test subjects' noses, they discovered that people with a rich network of friends (i.e., a wealth of social capital) were four times less likely to become ill compared with those with few friends.[28]

In the end, community and the social capital that it engenders may be humankind's last best hope, not just for happiness and the conviviality that it engenders, but, for surviving hard times.

It Takes Time—Not Money—to Create Community Social Capital

Time is money! We've all heard it many times. It's one of economism's mantras. The problem is that it's not actually true. Time isn't money; time is life! We live in time; we don't live in money.[29] In fact, it appears that our fixation with money is what leads us to experience time as scarce.[30] So if it is more life that we want, the answer may not be more money and the work that it demands, but a more spacious relationship with time. After all, what's extra money if you are stuck in traffic every day, perpetually sleep-deprived and, through it all, engage in work that fails to fill you with deep passion and purpose?

This ability to explore the tradeoffs surrounding time, money, and work seems to be taking hold because, increasingly, people are opting for more life (more time free from work) rather than more disposable income. In the United States, this shift was heralded in the 1990s, when Americans, for the first time, were saying that given the choice between a pay raise or more free time, they would choose more free time. And this wasn't just empty talk. For example, Boston College professor Juliet Schor found that millions of Americans were actually reducing the time they devoted to work in the first decade of the new millennium, either by cutting back on hours, working fewer jobs, or quitting work altogether. Forty-seven percent of these so-called downshifters cited job stress as the reason for pulling back; the rest were seeking more free time, more balance, and/or more meaning in their lives.[31] This trend appears to be continuing into the present.[32]

Whatever the specific rationale for downshifting, the big dividend is more free time. Specifically, time to spend with loved ones; time to learn new skills; time to devote to dormant passions; time to help friends and neighbors in need; not to mention time for satisfying household activities like cooking, growing food, home repairs, and child care—that is, jobs that downshifters previously used their hard-earned income to pay for. As Schor points out:

> All activities, whether they are monetized or not, have the potential to yield returns. We recognize wages and salaries as the returns to employment. But activities that do not earn dollars create returns as well. Doing work in one's own household, without a wage, is production. The cooked meal, the completed tax return, and the cared-for child all have economic value. . . . Even spending time with a friend, something one might consider purely uneconomic, strengthens social networks of support and reciprocity, which in turn creates access to resources.[33]

Downshifters have, in effect, a different understanding of what makes for a rich life. For them it's generally not about a newer car or a bigger house or more stuff. Rather than exhaust themselves seeking nonessential material remunerations, they opt to work fewer hours so that they have more time to devote to what seems to matter most in life—community, relationships, trust, joy, health, acts of creation, play and care giving. In the process, downshifters create social capital and do a good turn for planet Earth by reducing their ecological footprint. And downshifting doesn't appear to be a passing fad. In fact, Schor reports that, in her study, only 10 percent of people who opted to downshift regretted their choice; even those who did occasionally miss the money reported that they were happier than before.[34]

Community Solutions for Meeting Food and Energy Needs

As recently as one hundred fifty years ago, almost all the food that people ate came from the fields and pastures surrounding their communities. No more: These days, in the United States, the average food molecule travels some fifteen hundred miles before reaching people's plates. In some instances the distance is less (e.g., apples in season); in others far greater (e.g., dates from the Middle East). Given these distances, the current U.S. food system is anything but local!

Though it may be comforting to imagine your food coming from nearby family farms, those days are long gone. If you are buying your food at a supermarket, it comes to you thanks to the industrial food system—an impersonal approach to food creation entailing enormous land parcels; mechanized planting and harvesting; copious use of fertilizers and pesticides; extensive processing; and, yes, long-distance transport. All of this requires lots of energy. In fact, the U.S. food system runs a huge energy deficit. So huge that, on average, about ten times more energy (fossil fuels) is expended in the creation of the food Americans eat than the food itself actually contains.

Fortunately, there is hope. It comes in the form of local, organic, small-scale farming operations—the polar opposite of today's dominant big, distant, industrial, chemical food production model. Indeed, starting in the 1990s, Americans began to signal their disillusionment with industrial food, along with a growing enthusiasm for local, fresh, nonprocessed food; and farmers responded by creating a socioeconomic farming model dubbed "community-supported agriculture" (CSA for short). A typical CSA is composed of a collection of community members, say fifty households, that each pledge to support a local farmer. Each household makes an advance payment at the start of the growing season in return for a share of the anticipated farm harvest. These shareholders then receive fresh produce throughout the growing season. Early on, the weekly food boxes might be laden with lettuce, spinach, baby beets, turnips, radishes, scallions, snap peas; a bit later, in June, the food boxes contain chard, broccoli, zucchini, kale, and green beans, along with bouquets of flowers; then in July and August—in addition to much of the above—come such treats as carrots, tomatoes, sweet corn, cucumbers, peppers, and berries. Finally, at the end of the season, CSA members receive bounteous offerings of pumpkins, onions, squash, cabbage, parsnips, potatoes, among other delights. Locally raised eggs, honey, and meat also figure into many CSA operations.

CSAs are great for the community because residents receive fresh local produce each week. They make sense ecologically because they produce food using only about one-tenth as much energy per food unit as is the case in the far-flung industrial system. They are good for farmers because the growers receive money up front to purchase the seed and equipment they need to farm. And by cutting out middlemen in the form of food processors, packagers, transporters, and retailers, the big chunk of each dollar that normally would be lost from the local economy goes, instead, to compensating community farmers and even reducing food costs for CSA members.[35]

This is a win-win scenario that is catching on. The first CSA in the United States was established in the Northeast in 1984; today there are more than twenty-five hundred such operations—an average of fifty per state.[36]

If there still isn't a CSA near where you live, there is almost certainly a farmer's market. Their number, currently in excess of six thousand in the United States, has more than tripled since the mid-1990s, making them the fastest-growing component of the U.S. food economy.[37]

FARMER'S MARKETS AND COMMUNITY TIES

While fresh food is an important draw for farmer's markets, there is something more going on. To find out, a team of U.S. sociologists observed individual grocery store shoppers from the moment they entered the fluorescent-lit store aisles until they departed. The researchers then repeated their observation protocol at outdoor farmer's markets, again tracking shoppers from the moment they entered the farmer's market until they left. The result: Shoppers at farmer's markets had, on average, ten times more conversations than shoppers at supermarkets.[38] If you have had occasion to shop at a farmer's market, you probably won't find this surprising. Indeed, in contrast to the supermarket, where most people usually act as anonymous consumers, in the farmer's market they experience themselves as participants in a community, delighting in the colors, smells, and tastes of fresh food, while also conversing, face-to-face, with the actual people who grew the food that they will have for dinner.

Just as with food, citizens are also seeking ways to meet their energy needs in the places where they live. It won't be easy; creativity and conservation will be necessary, but it may just be possible. In fact, based on studies from the Institute for Local Self-Reliance, half of America's fifty states could meet all of their energy needs using renewable energy sources located within their borders, and most of the remaining states could meet a big percentage. For example, wind turbines and rooftop solar panels could supply more than four-fifths of New York's power needs. In other parts of the country the energy mix would differ: hydropower resources predominating in the Northwest; solar in the Southwest; and biofuels in the Southeast.[39]

Think of it as farmer's markets organized around electrons instead of food. Though it doesn't exist yet in the United States, such things have already been achieved in Europe and Canada through the creation of local wind power associations and energy cooperatives.[40] In this new paradigm, energy, rather than being understood as a commodity that people purchase from someplace far away, becomes something that cities and towns, and even individual citizens, help to produce and distribute to each other. For example, imagine the rooftop of your home equipped with solar panels or small wind turbines, allowing you to meet your personal electricity needs while also exporting your surplus electricity to neighbors.

Community Solutions for Meeting Shelter Needs

For almost our entire history as a species we humans have lived in small groups, seldom larger than a few hundred people. Even as recently as 1800 only three percent of the world's population lived in cities, the rest living in villages and towns. Thus, little in our social and evolutionary history prepares us for our current urban-suburban mode of living.[41] Perhaps because of this many people around the world are choosing to return to some semblance of village life. One manifestation of this is the growing prevalence of so-called ecovillages.

At present the Global Ecovillage Network connects an estimated fifteen hundred such village communities around the world. What they have in common is an intention to live lightly on Earth by sharing space, time, creativity, resources, and goodwill (all elements of social capital). Typically, ecovillage members live in small private dwellings, sharing things such as meeting rooms, play spaces, guest rooms, tool sheds, laundry rooms, workshops, and cars. Members—both adults and children—typically participate in the growing of food for communal consumption. Routine tasks like cooking and maintenance are also often shared. Decision making is democratic, often grounded in consensus. Dwellings are not spread out, as is typical in suburbia, but clustered together, connected by paths and walkways, with cars parked on the periphery of the village.

These ecovillages aren't designed and built by outside developers, but by the community members with an emphasis on simplicity and beauty, as well as ecological design principles that maximize energy efficiency and minimize waste. The net result of ecovillage living is, not surprisingly, a reduced per capita ecological footprint.

Of course, it is not always necessary to create a stand-alone ecovillage. The neighborhoods where we live *are*, in a sense, ecovillages in waiting—places where we can create genuine community by sharing and meeting each other's needs. In fact, this very idea has been seized upon by the Transition Towns movement, started in England in 2005 and now spread to eighty countries with more than one hundred participating towns and cities in the United States, including Los Angeles, Charlotte, Chicago, Pittsburgh, Milwaukee, and Tulsa.[42] Transition Town participants recognize that the time has come to *transition* away from our current growth-based, extractive fossil-fuel economy, toward life-serving, sustainable economies based on renewable resources and zero net growth. Such a transition entails literally reimagining our way of life in the realms of food, energy, health, waste, work, and habitation—tantamount to an economic, cultural, and spiritual renaissance.

Transition Towns participants acknowledge, "We truly don't know if this will work. Transition is a social experiment on a massive scale. What we are convinced of is this: If we wait for the governments, it'll be too late; if we act as individuals, it'll be too little; but if we act as communities, it might just be enough, just in time."[43]

Creating a Collaborative Economy— Opportunities for Participation

One outcome of our current impersonal economic system is that, if you are like most people, you pay money for almost everything that you need. You don't make it yourself or barter or share or lend; you buy it with money and then you own it. As a result, as Charles Eisenstein points out in his book, *Sacred Economics*, "we don't need each other." He has a point. Think about it: If you can pay money to acquire everything you need, you don't need the material support of those living in your neighborhood. If you need something, you simply pay cash for it. Done deal. Given this quandary, Eisenstein opines, "How can we build community when its building blocks—the things we do for each other—have all been monetized."[44]

No need to despair! Creative approaches to community-based economic life are bubbling up everywhere. Indeed, there is a quiet revolution brewing as people endeavor to create new economic models that genuinely enhance and enrich their lives. Take transportation. Cars can be very helpful at times, but owning one is expensive and maintaining one can be a hassle. But what if you could have the benefits of car use without the bother of owning one? Enter Zipcar, a community-oriented car-sharing company now operating in forty-nine U.S. cities with membership now doubling each year. In Boston alone there are more than 30,000 Zipcar users.

Take a minute now to imagine yourself as a Zipcar member freed from the hassle of car ownership. It's Tuesday morning and you need a car. No problem. You simply go to the Zipcar website on your computer (or tap the Zipcar app on your phone) and quickly reserve one of the cars located in your neighborhood. Done! By not having to bother with car purchase and upkeep Zipcar users save up to $600 per month in car-related expenses. The whole idea of Zipcar makes sense because cars, like many other things that Americans own, are seldom in actual use—e.g., the average privately owned car is only in use for one hour out of every twenty-four. In effect, every Zipcar vehicle on the road replaces seven to eight privately owned vehicles because by becoming car-sharers, members are able to get rid of their private cars.[45]

It's the same with power tools, lawn mowers, snow blowers, ladders, garden tools, and so on. These are all things that Americans buy but use only occasionally. For example, an estimated fifty million Americans own a power drill, and yet the average electric drill is used for less than ten minutes during its entire lifetime; the rest of the time it sits idle.[46] Clearly, it is not necessary for everybody on the block to own their own drill, hedge pruner, fence-post digger, and so forth. So, what if, just like a library for the sharing of books, there were neighborhood tool libraries? Actually, this need not be a "what if?" It's already happening in cities around the United States, such as Columbus, Atlanta, Berkeley, and Philadelphia. Again, pause to imagine this scene: It's Saturday morning and you have two home repair projects. Instead of driving out to Lowe's to buy needed tools, you simply walk over to your neighborhood tool library, check out the tools you will need, and you are all set. Tool libraries are a commonsense response to the recognition that we don't need to own products, since what we really want is access to the services that products provide.

Community-based exchange networks like Zipcar and tool libraries have the net effect of reducing consumption and waste. Another example is Craigslist—the well-known community marketplace and social hub. Go there to purchase used furniture, bicycles, appliances, you name it, not to mention to access job postings, lost-and-found listings, and discussion forum opportunities. This is the Internet acting as a force to enable people to decentralize, redistribute, and self-organize—in other words, to begin to take the economy back into their own hands.

Peer-to-peer renting is another example. Sites like RelayRides, Zilok, and Rentoid now act as platforms allowing citizens to loan things such as their personal cars, tools, cameras, and electronic equipment directly to each other. In a related development, citizens are even side-stepping the banking industry to extend cash loans directly to one another. Indeed, 10 percent of all loans in the United States are now made peer-to-peer, amounting to $6 billion a year. This makes sense insofar as it gets around the "asset managers, mortgage brokers, pension and mutual funds advisors, and the big banks themselves—who for the most part [have] introduced faceless transactions and overhead, while removing the community loyalty that [is] the glue in person-to-person lending."[47]

Freecycle is yet one more example of the emerging collaborative, community-based economy that models nature's ethic of mutual exchange. At present, Freecycle, operating in eighty-five countries, creates conditions (via the Internet) for people to arrange to give and receive "gifts." Grounded in the belief that "what's mine is yours," Freecycle is the perfect

PANERA BREAD GETS INTO THE ACT

What if you were told as you entered a restaurant to simply pay what you were able? Could a business, restaurant, or otherwise, survive with such a collaborative business strategy? Could people be trusted to pay their fair share? This is the question that Ron Shaich, director of Panera Bread Company, decided to explore in the Saint Louis suburb of Clayton. In Clayton's Panera there are no prices on the menu, only suggested "funding levels." If you have nothing, you don't pay; if you have cash you put it in a jar; if you have a credit card, you simply tell the cashier what you want to be billed, and that's that.

The experiment has been a success. The great majority of people (80 percent) pay the suggested amount or more. The message according to CEO Shaich is "to put faith in humanity. . . . People step up and they do the right thing." Folks who are unemployed or family members going through hard times benefit. One such beneficiary commented, "Sometimes I can pay full price, and sometimes I can't. It's nice to know you can come somewhere like this and they don't judge you. I'm very grateful for this place."

Based on the success of the Clayton experiment, Panera has extended the "pay-what-you-can" concept to cafes in Dearborn, Michigan, and Portland, Oregon, with more to follow.[48] Clearly Panera understands that the notion that people are inherently selfish and greedy is simply the cultural story of economism and not reality.

antidote to the greed-inducing belief that we live in a world of scarcity. One day, having more than you need—that is, living in a state of abundance—you might be a Freecycle giver; another day, in a time of need, you might be on the receiving end. Remarkably, each day, more than twenty-four thousand items, weighing approximately seven hundred tons, are exchanged through Freecycle.[49]

The more that things are shared, the less need there is to be manufacturing new things, which translates to reduced environmental impacts. It is estimated that for every pound of new product manufactured—whether it be smart phones or sofas—approximately thirty-two pounds of waste are produced.[50] Thirty-two pounds of waste for each pound of product! This means that if you use Freecycle to give your old sofa to someone else, not only would you be keeping approximately 100 pounds (the weight of the sofa) out of a landfill, but you'd also be eliminating the three thousand pounds of waste that would otherwise have been generated in the construction of a new sofa.[51]

In their own ways each of these new initiatives is the result of people seeing themselves as interdependent. This doesn't require a change in human nature—after all, we are inherently cooperative beings who seek community and relationship—but it does call for a change in the story that we choose to enact. And that change is already happening. All around us, a new story is being born. This story grows from the simple question, What can we do for each other? The answer, it appears, is *a lot*!

STEPPING BACK TO SEE THE BIG PICTURE

> Strange is our situation here upon Earth. Each of us comes for a short visit, not knowing why, yet sometimes seeming to [have] a divine purpose. From the standpoint of daily life, however, there is one thing we do know: that we are here for the sake of others.
>
> —Albert Einstein[52]

Humans are certainly capable of behaving in selfish and brutish ways at times, but, deep down, knitted into our biology, is an innate desire to meet each other's needs. Don't take my word for it. Simply call to mind something you have done in the last week that made life better for someone else. Take a minute. It need not have been anything grandiose. Now, as you recall that act of yours, register how you felt. If you are like the overwhelming majority of human beings, you felt good! It's essentially universal: Humans enjoy meeting each other's needs. In fact, there is probably nothing that gives us such deep and full satisfaction as helping each other.[53]

This raises the question, Why? Why do we take such pleasure in giving? Could it be that our predisposition to give grows from our innate gratitude for the gift of life that has been bestowed upon us? Consider: Simply by virtue of being born into this world each of us has received great gifts. After all, you didn't do anything to earn your life. In the moment of your birth, you received your life as a gift. And in those first days, months, and years of your life, there was someone there giving you the gift of care. This is no trivial matter. "Imagine finding yourself plunged into an alien world in which you were completely helpless, unable to feed or clothe yourself, unable to use your limbs, unable even to distinguish where your body ends and the world begins. Then, huge beings come and hold you, feed you, take care of you, love you. . . . Wouldn't you feel grateful?"[54]

Even in this very moment each of us is surrounded by gifts. Consider the air that you breathe. It is here for you, free for the taking—a gift. The same is true for water; you don't manufacture it every day; it is here for you, in abundance, as a gift. And it is no different with the food that you eat or the seeds that grow your food or the wood that created your house. Earth offers these things to each of us as gifts. After all, we didn't create planet Earth, much less ourselves, as is made clear in these words from Satish Kumar:

> We do not own our intellect, our creativity, or our skills. We have received them as a gift and grace. We pass them on as a gift and grace; it is like a river which keeps flowing. All the tributaries make the river great. We are [each one of us] the tributaries adding to the great river of time and culture, the river of humanity. If tributaries stop flowing into the river, if they become individualistic and egotistical, if they put terms and conditions before they join the rivers, they will dry and the rivers will dry too. To keep the rivers flowing, all tributaries have to join in with joy and without conditions. In the same way, all individual arts, crafts and other creative activities make up the river of humanity. We need not hold back, we need not block the flow. This is unconditional union. This is how society and civilization are replenished.[55]

Are you able to experience your life as a gift streaming through you—one small tributary in the great river of life—flowing to the future?

Insofar as we are able to recognize that we are recipients of myriad gifts, it is natural to feel gratitude. After all, what is gratitude but the recognition that we have received something and the concomitant desire to give something back in return?[56]

A Gift Economy

Though our dominant ideology—economism—engenders separation, and with it environmental and social distress, it is not the only option open to us. Indeed, we can, if we choose, devise other social transaction systems to meet our daily needs. For example, many people down through the ages have chosen to live in so-called gift economies grounded in the understanding that human beings live in a world of abundance, not scarcity. This being so, there is no need to hoard. For example, if you have twenty ears of corn and you only need two, why not give the other eighteen away? Then, the recipients of your corn will experience gratitude and, naturally, will want to give back in return. Indeed, the more we give, the more gratitude we will engender among those in our community.

Through giving, we, in effect, create an empty space—a vacuum—that tugs upon the whole, ensuring that gifts flow back to us, as necessary. So it is that in a gift economy, paradoxical as it may seem, security actually comes from giving! Suddenly, life is no longer about how much you keep but how much you give. This turns our contemporary notions of wealth upside down.[57]

Receiving is just as important as giving in a gift economy, for by graciously receiving others' gifts, we submit to the gift. Eisenstein puts it this way:

> Receiving and giving go hand in hand. . . . To give and to receive, to owe and be owed, to depend on others and be depended on—this is being fully alive. To neither give nor receive, but to pay for everything; to never depend on anyone, but to be financially independent, to not be bound to a community or place, but to be mobile . . . such is the illusory paradise of the discrete and separate self. Independence is a delusion. The truth is, has always been, and always will be that we are utterly and hopelessly dependent on each other and on nature.[58]

Though the idea of a gift economy may sound like pie-in-the-sky idealism, it appears to be the way humans are wired to behave and respond. In fact, just since the 1990s almost all of us have become participants in a gift economy known as the Internet. For example, the free online encyclopedia, Wikipedia, is the result of tens of thousands of people giving their time to create tens of millions of articles of information that you and I have free access to. It's the same with Internet open-source software and file-sharing provisions that allow people to freely share news, videos, insights, and more with each other, alternately serving as both givers and receivers. The fact that so many people devote their time and energy to these projects—not for monetary gain, but because they find it satisfying to create something that other people can use—is yet another illustration of how the assumptions of economism—especially that the profit motive is the only thing keeping us from being lazy louses—is an illusion.

None of this is to deny the utility and effectiveness of money as a means of exchange. Rather, the point is simply that by limiting ourselves exclusively to monetary transactions we miss out on myriad opportunities to create the critical social capital that emerges as we cultivate local gift economies.

We Are Here to Give Our Gifts

What if our calling as human beings is not to become servants of economism, working ourselves silly so that we can fill ourselves with stuff that may ultimately leave us more bereft than happy? Rather, what if we have been born solely and simply to give our gifts? What if that's really why we are here? In this context, what if you understood that your gift is anything that allows you to give and connect to Earth's community of life? Bill Plotkin puts it this way:

> The gift you carry for others is not an attempt to save the world but to fully belong to it. . . . Discovering your unique gift to bring to your community is your greatest opportunity and challenge. The offering of that gift—your true self—is the most you can do to love and serve the world. And it is all the world needs.[59]

WE ALL HAVE A GIFT TO GIVE

When you were growing up, did anyone ever ask you what your gift was— what you were born to give to the world? If not, I ask you now, What is your gift? What is the beautiful thing that you are here to give? It need not be grandiose; it just needs to be yours to give. In this vein, Hawaiian poet and community organizer Puanani Burgess tells of a time when she asked a group of high school students to each tell three stories—the story of their name, the story of their community, and the story of their gift. It was going well until Burgess came to a certain young man. He did fine with the story of his name and his community but when it came time to tell the story of his gift, he bristled: "What, Miss? What kind gift you think I get, eh? I stay in this special-ed class and I get a hard time read and I cannot do that math. And why you make me shame for, ask me that kind question, 'What kind gift I have?' If I had gift, you think I be here?"

In that moment Burgess felt awful for shaming the young man as she had. Then, two weeks later she was at the local grocery store and saw that same young man down one of the aisles. Still feeling ashamed, she tried to flee but he caught up with her and said: "You know, I've been thinking, thinking, thinking. I cannot do that math stuff and I cannot read so good, but Aunty, when I stay in the ocean, I can call the fish and the fish he come, every time. Every time I can put food on my family table. Every time. And sometimes when I stay in the ocean and the Shark he come, and he look at me and I look at him and I tell him, 'Uncle I not going take plenty fish. I just going to take one, two fish, just for my family. All the rest I leave for you.' And so the Shark he say, 'Oh, you cool, brother.' And I tell the Shark, 'Uncle you cool.' And the Shark, he go his way and I go my way."

This is a story of a young man with a remarkable gift, living in a culture that fails to recognize that gift. It need not be this way. Imagine schools centered on a gift-based curriculum—where the focus is on bringing forth and deepening each child's gifts.[60]

Growing up most of us were encouraged to follow society's script. Some rebelled and forged a path uniquely their own, but, sadly, most of us fell in line, meeting society's often stifling expectations rather than our own. This works for a time, but eventually a day comes—maybe when we are forty—when we wake up and feel an overwhelming emptiness. When we mention this to others they say, "Oh, it's just a midlife crisis; you'll get over it." What it really is, of course, is our long-dormant soul awakening us to the lie that has become our life. It is only when we realize that life itself is a gift and that we are here to give of ourselves that we become free. After all, what we take in this life dies with us. Only our gifts live on.[61]

WRAP-UP: EXPANDING ECOLOGICAL CONSCIOUSNESS

> I don't know, of course, what's going to happen, but it seems to me imaginable that a time could come when we will either have to achieve community or die, learn to love one another or die. We're rapidly coming to the time, I think, when the great centralized powers are not going to be able to do for us what we need to have done. Community will start again when people begin to do necessary things for each other.
>
> —Wendell Berry[62]

We could easily look at all the change, upheaval, and disturbance occurring throughout the world today and conclude that things are falling apart and that the human project is headed for disaster. However, scientists have learned that, in the natural world, major shifts are often preceded by great turmoil, chaos even. For example, Ilya Prigogine won the Nobel Prize in 1977 by demonstrating that certain chemical systems actually shift to greater order after they are disturbed. In other words, disorder can actually act as an ally, cajoling a system to self-organize into a more ordered state, one better attuned to the demands of its changed environment.[63]

The idea that greater order can emerge from states of chaos appears counterintuitive until we think about it from the perspective of our own life experience. After all, the times of hardship and suffering—when we personally descend into chaos—often lead to personal growth and increased integrity. It is the same with social systems: disruption, confusion, and chaos are often necessary to awaken creativity and move to a higher order. Seen in this light, the very unraveling of our old worldview—a view based on separation from nature, economic exploitation of Earth's body, and dominance relations among people—may very well be providing the creativity and energy necessary to catalyze the birthing of a new story. As Margaret Wheatley, author of *Leadership and the New Science*, astutely observes:

> In the dream of dominion over all nature, we believed we could eliminate chaos from life. We believed there were straight lines to the top. If we set a goal or claimed a vision, we would get there, never looking back, never forced to descend into confusion or despair. These beliefs led us far from life, far from the processes by which newness is created. And it is only now, as modern life grows ever more turbulent and control slips away, that we are willing again to contemplate chaos. . . . The destruction created by chaos is necessary for the creation of anything new.[64]

A new worldview won't take hold because of clever arguments or impassioned pleas; it will only emerge after many, many people come to understand that the present worldview (story) is causing irrevocable damage and must be abandoned. To paraphrase Einstein, a new story begins to gain traction when people come to understand that the problems they face can no longer be solved from the same level of consciousness that created those problems in the first place.

So, the first step is to wake up from the self-defeating belief that we are powerless to effect change in our lives and beyond. This tragic misunderstanding of self is illustrated in a story about Mohini, a regal white tiger who once lived at the National Zoo in Washington, DC. For many years Mohini paced relentlessly back and forth in her twelve-foot-square cage. Then, the zoo staff decided she deserved something better and set about to create an outdoor park for her, composed of hills and trees and a pond. However, when Mohini was released into her new home, she retreated to a secluded corner of the compound where she spent the rest of her life, pacing back and forth until she had worn an area twelve feet square, bare of grass.

Tara Brach, who tells Mohini's story in her book, *Radical Acceptance*, remarks: "Perhaps the biggest tragedy in our lives is that freedom is possible, yet we can pass our years trapped in the same old patterns. . . . Like Mohini, we grow incapable of accessing the freedom and peace that are our birthright."[65]

But, of course, this need not be. The time has come to wake up. As we do this, we create a new story of self, finally grasping that our purpose—the reason we are here—is to give the unique gift that is ours to give in the service of life. Also, we create a new story of the *Other*, coming to know that more for you is more for me and that *belonging* is what matters, not *belongings*. And, finally we create a new story of Earth, delighting in the realization that Earth is our larger body, the sacred whole that we dwell within.

APPLICATIONS AND PRACTICES:
BIRTHING AWARENESS

> Imagine yourself for a moment living in a more mindful and attentive way. See yourself moving more slowly through life, taking time to notice things, time to find your balance as you walk, time to notice how things look and smell around you. See yourself looking deeply into other people's eyes as you talk with them, studying their faces with attention and sensitivity. See yourself deeply enjoying the pleasure of love-making, a fresh salad, a starry evening sky, walking barefoot in wet grass. Imagine yourself gazing steadily inward, knowing and accepting yourself, your feelings, longings, spiritual intuitions, dreams. See yourself so clear, so centered and strong, that you can calmly remain with those in crisis, listening, loving, understanding, giving.
>
> —Marc Burch[66]

If we humans survive the convergence of crises that now envelop us, it will be because we have gained the capacity to see with the wide-open eyes of relationship, not the armored eyes of separation. There is hope. Indeed, these days, people all over the world are waking up as they seek a new understanding of what it means to be truly and fully human in these challenging times.

Awakening by Cultivating Mindfulness

A particularly powerful awakening practice is called "mindfulness meditation." Variations of this practice are now being taught to people in all walks of life—business personnel, teachers, professional athletes, students—for the purposes of increasing concentration, insight, and overall awareness.

In its most basic form, mindfulness meditation consists of sitting quietly while placing full attention on one's breathing—noticing the incoming and outgoing breath—without verbalizing or conceptualizing. The idea is that by using the breath to bring attention to the *here and now*—the present moment—it is possible to extricate ourselves from the morass of mental images that often both flood and cloud our awareness.

Although it may sound easy, mindfulness meditation can be quite challenging, especially at first. For example, often our minds introduce a juicy thought before we have even completed one in-breath; or if we do manage to complete a single breathing cycle, our mind jumps in to congratulate us.

If you doubt this, pause for a few moments now and give mindfulness meditation a go. Begin by assuming an alert but relaxed sitting position.

Cast your eyes downward and then, when you are ready, bring your attention to your breathing—in breath . . . out breath—simply witness your breathing for five minutes.

As you engage in this simple practice you will discover, through direct experience, that the human mind tends to be a very busy place—our thoughts wander hither and yon. The value of meditation is that it allows us to bring awareness to our mind's antics, and with this awareness comes understanding and, given sufficient practice, genuine freedom from conditioned ways of reacting, seeing, and being.

When you find yourself sidetracked in thought during meditation, you can assume the role of a detached observer, simply watching your thoughts go by without getting caught up in them. Welcome each new thought, acknowledge it, and then let it pass away of its own accord. For example, if you get caught up in fretting, simply note, "worry mind." Do the same for "planning mind," "judging mind," "fearful mind," and so forth.

As you explore mindfulness meditation, it may help to picture yourself as a rock resting by a riverbank and to think of your thoughts as the water flowing down the river; just let those thoughts flow by. Normally, we engage with our passing thoughts and get caught up in their emotional charge—that is, we *become* our thoughts. "In our so-called normal state of consciousness we are therefore continually lost in the dramas of our lives, unaware of how the process that creates our stories is taking place."[67]

Meditation teacher Wes Nisker uses a movie theater analogy to explain how mindfulness meditation works:

> When we are watching the screen, we are absorbed in the momentum of the story, our thoughts and emotions manipulated by the images we are seeing. But if just for a moment we were to turn around and look toward the back of the theater at the projector, we would see how these images are being produced. We would recognize that what we are lost in is nothing more than flickering beams of light. Although we might be able to turn back and lose ourselves once again in the movie, its power over us would be diminished. The illusion-maker has been seen.
>
> Similarly, in mindfulness meditation, we look deeply into our own movie-making process. We see the mechanics of how our personal story gets created, and how we project that story onto everything we see, hear, taste, smell, think, and do.[68]

In sum, the practice of mindfulness meditation is a lifelong discipline that leads to ever-greater presence and ever-widening consciousness. With practice, it has the potential to catalyze both personal and global awakening.

Awakening by Doing

Just as quieting and turning inward are important for awakening, so too, in equal measure, is doing—that is, taking concrete actions to contribute to the birthing of a new story grounded in kinship, cooperation, and peace. There are hundreds of ways to participate in this cultural reformation. Here are three for illustrative purposes.

i. Birth a Gift Circle

Explore the sensibility and delight of the gift economy by creating a gift circle. It is easy to do. Just gather with some friends and neighbors—ten is a good number. Then, invite everyone to take a turn sharing a need that they have. It could be anything. For example, maybe you, yourself, need someone with a strong back to help you put in a garden bed or build a stone wall; or perhaps you need someone with technical skills to help you solve a computer problem; or maybe you would like someone to lend you an electric drill or a pasta maker or a ladder. After each person speaks, others in the circle can, if they choose, offer to meet the need or offer ideas for how it could be met.

Then comes a second round where each person has a turn to tell about something they have to give. It could be something that is no longer needed like a toaster oven or a chair or art supplies; alternatively, it could be a willingness to lend tools or to offer assistance with bike repair. Here, too, as participants offer their gifts, anyone can simply say, "I would like that" or give the name of a person or organization that might benefit from the gift being offered.

A gift circle concludes, quite naturally, with expressions of gratitude both for the tangible expression of needs and gifts and the pleasure of each other's company. The following day it is customary for the convener to send an e-mail summarizing each person's needs and gifts, along with contact information.

The profundity of the gift circle is that it pares the human economy down to its essentials—namely: needs, gifts, gratitude. Regular participants discover that they don't need to spend as much time working to earn money simply because some needs that they previously met using earned income are now being met by those in their gift circle. And once a culture of gifting begins to take root, it has a way of building on itself:

Gifting sets in motion a cycle of generosity where one gift prompts another.
As generosity grows, emotions that destroy relationships, especially jealousy

and competition, recede. . . . In a generous society, we don't need to reduce others in order to feel superior. We start to enjoy each other, even encourage each other. Once we experience this quality of community, it's hard to go back to pettiness and jealously. We're so much more gifted, all of us, in a generous society.[69]

Citizens don't have to wait for government leaders to create a gift economy; it doesn't happen that way. It's enough to simply live in the spirit of the gift by enhancing beauty and creativity rather than diminishing it; sharing wealth rather than amassing it; reinforcing the cyclical processes of nature rather than violating them; and freely offering our gifts, trusting that what goes around, comes around.[70]

ii. Birth Personal Competency

One of the ironies of contemporary education is that spending more years in school doesn't necessarily mean the acquisition of more practical life skills. For example, college graduates often have limited skills when it comes to building things, making and mending clothes, growing food, repairing things that break down (e.g., appliances, cars, leaking pipes), and so forth. One who is lacking in such basic construction and repair skills is largely dependent on others for his/her basic survival. Of course, this need not be so. Each of us can begin—literally—to take things into our own hands.

It's high time. Start by simply giving some thought to something you would like to have for yourself and, instead of buying it, make it! That's right: Use your hands, materials that are lying around, and your innate creativity to make something meaningful for yourself—something that will last and that has some utility. And while you are at it, consider repairing something. Pick something that is broken and that you would normally just throw away; but, now, instead of tossing it, fix it so that you can continue to use it over and over into the future.

Finally, in this same spirit of birthing personal competency, go to a nearby recycling bin and fish out something that attracts you. Then, rather than allowing this object to be downcycled, use your innate creativity to upcycle it into something new and useful.[71] By increasing your skill in these practical ways, you will also be improving your gift-giving capacity.

iii. Birth Something Novel in Your Neighborhood

As Jay Walljasper of the Project for Public Spaces points out, "The neighborhood is the basic unit of human civilization. Unlike cities, counties, wards, townships, enterprise zones, and other artificial entities, the neighborhood is easily recognized as a real place. It's the spot on Earth we call home."[72]

So, consider how it would be if we joined with the folks in our neighborhoods to consider the question, What's possible here? We might start by simply offering hospitality, good conversation, acts of kindness, and trust to each other. And in this spirit of conviviality, what if we removed the fences separating properties and refashioned them into benches for sitting and swings for playing? And how about if we envisioned neighborhood intersections—not as simply places where two streets come together, but—as neighborhood gathering places with bulletin boards, food stands, tables, play spaces, and art?

And why stop there? Imagine creating neighborhood enterprises—for example, producing bread in wood-fired bread ovens; exchanging home-grown vegetables, herbs, and eggs from backyard gardens and hen houses; displaying handmade objects for barter or sale, growing u-pick strawberries and blueberries next to our sidewalks; offering free foot massages to people trekking home from work; placing freecycle items along with favorite books in special *giving boxes* by our mailboxes. And while we are at it, how about making a map of neighborhood households based on an inventory of household gifts and talents, as well as needs? Yes, there are a thousand ways to rekindle our humanity, and most of them don't require much, if any, money.

READY, GET SET, GO!

Here's a recipe guaranteed to produce something useful for your neighborhood while engendering creativity and connection. The objective is simple—join with others to give a gift to your neighborhood. Here are four provisos that will ensure your success. First, whatever you build must be simple and accessible. Second, it can't cost any money—this means that any tools or materials will have to be acquired without the exchange of money. Third, it has to be handmade and created together with others in your neighborhood. And, fourth, it has to provide something useful for your neighborhood.[73]

In sum, when it comes to creating community, it's not necessary that you have a grand plan; it's enough to simply begin—to take that first step. Once you do that, you can be sure that forces, both within and outside you, will come to your aid and guide you forward.

QUESTIONS FOR REFLECTION

- What new story about who you are and why you have been born are you ready to birth? What "labor" would be required to birth this new story of self?

- Imagine that starting tomorrow you have four extra hours each day to spend in any way you would like. What would you do with this gift of time? Make a list. Now imagine the same scenario, but this time you have four fewer hours each day. What would you eliminate from your life? Again, make a list. Finally, considering both lists, what insights and ideas emerge?

- What would you have to do to meet all your basic needs for food, clothing, shelter, and transportation within a ten-mile radius of where you live?

- How would our societal conception and use of money change if we were to enact the belief that wealth results from how much we give, not how much we keep?

- What is a gift that you possess that you have not fully acknowledged? What are the consequences of holding your gift in exile?

- What's a "door" in your life that you are afraid to open? What's your story about what's on the other side of that door? What's a new story that would allow you to open this door?

EPILOGUE

Mis estimados: Do not lose heart. We were made for these times.

—Clarissa Pinkola Estes[1]

Not long ago, I heard high-pitched shrieking coming from my backyard. My six-year-old daughter, Katie, was out playing with her friend Olivia. I rushed to the back porch and implored: "What is it that has you screaming so?" Breathless, Katie exclaimed, "A monarch; we just saw a monarch butterfly!" Ah, to see with fresh eyes the wonder of a butterfly!

Biologist Elisabet Sahtouris has suggested that the process of metamorphosis whereby an engorged caterpillar transforms into a magnificent butterfly is an appropriate metaphor for what is now happening on planet Earth.

Think about it: Caterpillars are prodigious consumers, able to eat many times their body weight in a single day and, in the process, to sometimes destroy their host plants. It seems that we humans, given our own prodigious consumption, have been acting like the proverbial caterpillar, gobbling up everything in sight.

There is more to this comparison. When a caterpillar is so stuffed it can hardly move, it attaches itself to a branch and forms a hard shell around its body. Then, the cells of the caterpillar's body begin to break down, forming a kind of soup. But not all of the cells dissolve. Some that have lain dormant in pockets in the caterpillar's skin suddenly become active. Remarkably, these surviving caterpillar cells contain, within their DNA, the instructions for creating a butterfly. Because these cells can "imagine" a different way of being and living, biologists refer to them as "imaginal cells."

We humans are in need of a transformation along the lines of that achieved by the caterpillar. Our consumption-based lifestyle is devastating Earth and leaving us bloated. Fortunately, like the caterpillar, we also have imaginal cells in the form of humans who hold an image—a vision—for an entirely new way of being and seeing and living.[2]

In this vein, I draw hope from the story of the poet Robert Desnos, a prisoner in a Nazi concentration camp during World War II. One day Desnos, along with others, was loaded onto a flatbed truck. They all knew that the truck was going to the gas chamber. No one spoke. When they arrived, the guards ordered them off the truck. But then, as they began to move toward the gas chamber, Desnos suddenly jumped out of line and grabbed the hand of the woman in front of him and began to read her palm. The forecast was good: A long life, many grandchildren, abundant joy. A person nearby offered his palm to Desnos. Again, Desnos foresaw a long life filled with happiness and success. The other prisoners became animated, eagerly thrusting their palms toward Desnos, and, in each case, he foresaw a long and joyous life. The guards became visibly disoriented. Minutes before, they were confident in their mission, the outcome of which seemed inevitable; but now they became tentative in their movements. In fact, Desnos was so effective in creating a new reality that the guards eventually ordered the prisoners back onto the truck and took them back to their barracks.[3] In this story Desnos acted as a kind of imaginal cell—someone mired in a soup of death, but with the ability to imagine something different!

What about you? Can you conjure the world you long for? You might give it a try by completing the open sentence, *I want a world where . . .* When I invite college students to do this, I hear things like:

- I want a world where there are no weapons, no wars, no hatred.
- I want a world where children are cherished and nurtured and where elders are honored.
- I want a world where everybody is *family* and where people share, rather than hoard.

Many today seem to long for a kinder, gentler world—a world infused with care and compassion and honesty. What is the world that you long for? What would be removed? Added? Can you paint it? Can you tell how two strangers would greet each other? Place your hand on your forehead—the locus of cognition—and describe the world that you long for. Now hold your hand over your heart—the seat of passion and desire—and ask, What is the world I long for? Do you have the courage to imagine it, the will to speak of it, the determination to play your part in manifesting it? Here is a vision and promise to sit with:

> By the end of your lifetime you will be living in a world unimaginably more beautiful than the one you were born into. And it will be a world that is palpably improving year after year. . . . Prisons will no longer exist and violence will be a rarity. Work will be about, "How may I best give of my gifts," instead of "How can I make a living?" Crossing a national border will be an experience of being welcomed, not examined. Mines and quarries will barely exist, as we reuse the vast accumulation of materials from the industrial age. We will live in dwellings that are extensions of ourselves, eat food grown by people who know us and use articles that are the best that people in the full flow of their talents [can] make. . . . We will live in a richness of intimacy and community that hardly exists today. . . . And most of the time the loudest noises we hear will be the sounds of nature and the laughter of children.
>
> Fantastical? The mind is afraid to hope for anything too good. If this description evokes anger, despair, or grief, then it has touched our common wound, the wound of separation. Yet the knowledge of what is possible lives on inside each of us, inextinguishable.[4]

We are at the time of "The Great Turning." It is an in-between time. We may not make it; the dangers are real and daunting. Humanity has never experienced such a perilous time; we live in the shadow of nuclear annihilation, widespread chemical poisoning, catastrophic climate change, resource wars, epidemic disease outbreaks, and more—a cacophony of threats and wounds created by an upstart species, *Homo sapiens*—us!

It is tempting to conclude that, as a species, we have made a colossal mistake, fixated as we now are on constant growth and consumption. But maybe, just maybe, it will turn out not to have been a wrong turn after all; maybe the wacky frenzy of growth of the last two centuries with all its inflicted wounds has been necessary to get us to where we are going. Maybe it's a kind of developmental stage that our species has had to go through in order to achieve greater maturity. Indeed, in many respects we are still a species in its adolescence. For example, like a human adolescent, our present-day culture has an enormous appetite and is constantly downing resources. What's more, just like a teenager, contemporary culture is often self-centered, self-indulgent, wild, reckless, and impatient—in constant pursuit of action, power, and novelty.

And just like a child who grows and grows, year after year, all the while taking and taking from his parents, so it is that we humans have been growing and growing, taking and taking from Earth! But, of course, at some point the human adolescent stops growing, maybe it's at five feet, maybe at six, but growth does stop. This happens naturally. With the cessation of growth, the young person begins to assume the qualities and perspectives of genuine adulthood, gradually coming to realize that one's purpose is to recognize and meet the needs of others by sharing one's particular gifts.[5]

Historically, this transformation from adolescent to adult was facilitated through a process of initiation where the young person, with the help of elders, was subjected to a series of trials and tribulations designed to catalyze a profound awakening. Often (especially in the case of males) the youth was sent away and directed into an experience—often involving fasting and hardship—that shattered his habituated modes of understanding and knowing. In effect, the adolescent's world fell apart, and he didn't know who he was anymore. This temporary dissolution of the ego helped the youth begin to understand his place among his people. If he survived (and survival was not guaranteed), he returned as an adult, bearing a gift for his people.

I believe that humankind is now in the midst of just this kind of ordeal as we struggle to birth a new understanding of the human. The amalgam of planetary crises that humankind now faces can be seen as a kind of "coming of age" ordeal inviting us to finally grow up as a species.

A Great Turning is being enacted as we turn away from our old, adolescence-imbued story of the separate self that exists to take, exploit, and demand; and turn to the new story of the generative adult who sees with the new eyes of kinship, compassion, and generosity. In this new story, we are no longer *apart from* but, instead, *a part of* the family of life on Earth.

There is more to this metaphor. About the time teenagers stop growing, they often come to see their parents with new eyes. Recognizing all that their parents have given them—how much they have received from them—the teenager experiences gratitude and a growing readiness to be a giving (not solely taking) member of his or her family. This is also the time that the young person is likely to fall in love for the first time. Filled with fervor, the young lover wants to give gifts to his or her beloved and to cocreate something in partnership. This very same dynamic now appears to be playing out for humankind as a whole. Like the young human filled with love for a parent or mate, a deep desire is rising in humankind to reciprocate—to give back to Earth—for the gifts of life and nurturance that "she" has bestowed upon us. We know when this begins to happen because things that once escaped our notice are now felt deeply—for example, a year ago we may have heard about a fish kill in a nearby stream and thought little of it; now, we are profoundly saddened, knowing that a part of our larger body is afflicted.[6]

In sum, our situation today as humans is analogous to a birth crisis. The baby is breach—caught sideways in her mother's womb. The mother is in agony. Lots of folks are there trying to help. Time is running out . . . fear is rising . . . the mother is sweating terribly . . . there are tears . . . and then the baby shifts and becomes unstuck. No one knows how or why or what caused the baby to shift. The shift wasn't engineered; it just seemed to happen; call it "life longing for life." And out comes a baby—the most beautiful baby anyone has ever seen. But, in truth, the baby is no more beautiful than other newborns. It is the adults' way of seeing that has shifted; they can now see the sacred in everything, including each other. This seeming miracle happens not because an intervention of law or science or technology or economics saves the child but because the alchemy of hardship and suffering softens us and opens our hearts, reminding us of what is precious and true. Yes, it is upheaval and tumult that grow us up. Birthing a new self, a new culture, a new world will be terribly difficult and demanding, but the result—the full-fledged flowering of our shared humanity—is certainly worth the effort.

NOTES

PREFACE TO NEW EDITION

1. Vaclav Havel, quoted from Sharif Abdullah, *Creating a World That Works for All* (San Francisco: Berrett-Koehler, 1999), viii.

PART I: EARTH, OUR HOME

1. Matthew Fox, *Coming of the Cosmic Christ* (San Francisco: Harper San Francisco, 1988), 32.

2. Steve VanMatre, *Earth Education* (Greenville, WV: Institute for Earth Education, 1990). Contact information: Institute for Earth Education, Cedar Cove, Greenville, WV 24945.

CHAPTER 1: DISCOVERY

1. Thomas Berry, *The Great Work* (New York: Bell Tower, 1999), 14–15.

2. Chet Raymo, *Skeptics and True Believers* (New York: Walker, 1998), 243.

3. Michael J. Gelb, *Discover Your Genius* (New York: HarperCollins, 2002).

4. Paul Krafel, *Seeing Nature: Deliberate Encounters with the Visible World* (White River Junction, VT: Chelsea Green, 1999).

5. Krafel, *Seeing Nature*, 16.

6. Brian Swimme, *The Hidden Heart of the Cosmos* (Maryknoll, NY: Orbis Books, 1996).

7. Michael Seeds, *Horizons: Exploring the Universe* (Pacific Grove, CA: Brooks/Cole, 2002); Chet Raymo, *An Intimate Look at the Night Sky* (New York: Walker, 2001).

8. Inspired by Johns Hopkins geologist George Fisher.

9. Raymo, *An Intimate Look*.

10. Phillip Ball, *A Biography of Water: Life's Matrix* (Berkeley: University of California Press, 2001); Seeds, *Horizons*.

11. Brad Lemley, "The Last Great Mystery," *Discover* 23, no. 4 (2002): 32–39.

12. Raymo, *An Intimate Look*, 208.

13. Lemley, "The Last Great Mystery."

14. Brian Swimme and Mary Evelyn Tucker, *Journey of the Universe* (New Haven, CT: Yale University Press, 2011), 36.

15. Chet Raymo, *Natural Prayers* (St. Paul, MN: Hungry Mind Press, 1999).

16. Richard Heinberg, *The End of Growth* (Gabriola Island, BC: New Society Publishers, 2011).

17. Lynn Margulis and Dorion Sagan, *Microcosmos* (Berkeley: University of California Press, 1997).

18. David Suzuki, *The Sacred Balance* (Amherst, NY: Prometheus Books, 1998).

19. Suzuki, *The Sacred Balance*; Seeds, *Horizons*.

20. Lynn Margulis, *Symbiotic Planet* (New York: Basic Books, 1998), 71–72.

21. Genesis 1:3.

22. Andrew Beattie and Paul Ehrlich, *Wild Solutions* (New Haven, CT: Yale University Press, 2001), 32–33.

23. Deepak Chopra, *The Book of Secrets* (New York: Harmony Books, 2004), 186.

24. Swimme and Tucker, *Journey of the Universe*, 8–9.

25. Brian Swimme, *The Universe Is a Green Dragon* (Santa Fe, NM: Bear, 1985).

26. Elisabet Sahtouris, *EarthDance: Living Systems in Evolution* (Lincoln, NE: iUniversity Press, 2000) 78.

27. Christopher Uhl and Dana Stuchul, *Teaching as if Life Matters: The Promise of a New Education Culture* (Baltimore: Johns Hopkins University Press, 2011).

28. Diarmuid O'Murchu, *Our World in Transition* (New York: Crossroad, 1992).

29. Thomas Berry, *The Dream of the Earth* (San Francisco: Sierra Club Books, 1988), 123–24.

30. Swimme and Tucker, *Journey of the Universe*, 2.

31. This figure was inspired by a sketch in Seeds, *Horizons*, 231.

32. Swimme, *The Hidden Heart of the Cosmos*, 52–53.

33. Raymo, *An Intimate Look*, 6, 8.

34. Thomas Stella, *The God Instinct* (Notre Dame, IN: Sorin Books, 2001).

35. Swimme, *The Hidden Heart of the Cosmos*, 43.

CHAPTER 2: COMING TO AWARENESS

1. David James Duncan, *My Story as Told by Water* (San Francisco: Sierra Club Books, 2001), 185.

2. Thomas Berry, "The Ecozoic Era," accessed at www.lightparty.com/Visionary/EcozoicEra.html.

3. David Abram, *Becoming Animal: An Earthly Cosmology* (New York: Pantheon Books, 2010), 59.

4. Abram, *Becoming Animal.*

5. Abram, *Becoming Animal*, 101.

6. The mosquito scenario is inspired by Abram, *Becoming Animal.*

7. Abram, *Becoming Animal*, 62.

8. Joanna Macy and Molly Young Brown, *Coming Back to Life: Practices to Reconnect Our Lives, Our World* (Gabriola Island, BC: New Society Publishers, 1998).

9. Abram, *Becoming Animal*, 73, 77, 78.

10. This figure was inspired by Tyler Volk, *Gaia's Body* (New York: Copernicus/Springer-Verlag, 1998), 120.

11. Volk, *Gaia's Body.*

12. Brian Swimme, *The Universe Is a Green Dragon* (Santa Fe, NM: Bear, 1984), 37.

13. Volk, *Gaia's Body*, 124.

14. Paul Krafel, *Seeing Nature: Deliberate Encounters with the Visible World* (White River Junction, VT: Chelsea Green, 1999).

15. Volk, *Gaia's Body*, 128–29.

16. Evan Eisenberg, *The Ecology of Eden* (New York: Alfred A. Knopf, 1998), 23.

17. Eisenberg, *The Ecology of Eden.*

18. Krafel, *Seeing Nature.*

19. Krafel, *Seeing Nature.*

20. David Suzuki, *The Sacred Balance* (Amherst, NY: Prometheus Books, 1998), 8–9.

21. Suzuki, *The Sacred Balance.*

22. Frank H. Bormann and Gene E. Likens, "Nutrient Cycling," *Science* 155 (1967): 424–29.

23. Mary Guterson, "Living Machines," *In Context* 35 (1993): 37–38.

24. Guterson, "Living Machines."

25. Sim Van der Ryn and Stuart Cowan, *Ecological Design* (Washington, DC: Island Press, 1996).

26. For more ideas, see: http://toponlineengineeringdegree.com/?page_id=116 and http://pinterest.com/vluftensteiner/recycling-and-upcycling.

27. Thomas Berry, *The Dream of the Earth* (San Francisco: Sierra Club Books, 1988), 171.

28. Richard Louv, *Last Child Left in the Woods* (New York: Workman Publishing Company, 2005).

29. Chellis Glendinning, "Technology, Trauma, and the Wild," in *Ecopsychology: Restoring the Earth, Healing the Mind*, ed. Theodore Rozak, Mary Gomes, and Allen Kanner (San Francisco: Sierra Club Books, 1995), 41.

30. Cited in Louv, *Last Child Left in the Woods*.

31. Louv, *Last Child Left in the Woods*.

32. Linnie M. Wolfe, ed., *John of the Mountains: The Unpublished Journals of John Muir* (Madison: University of Wisconsin Press, 1938).

33. Duncan, *My Story as Told by Water*, 185.

34. Thich Nhat Hanh, *The Heart of Understanding* (Berkeley, CA: Parallax Press, 1988), 21.

35. I was first introduced to this practice through Project NatureConnect (www.ecopsych.com).

36. Jon Kabat-Zinn, *Full Catastrophe Living* (New York: Dell, 1990), 27–28.

37. Charles Eisenstein, *The Yoga of Eating* (Washington, DC: New Trends, 2003).

38. Steve Van Matre, *Earth Education: A New Beginning* (Greenville, WV: Institute for Earth Education, 1990).

CHAPTER 3: CULTIVATING COMMUNITY

1. Derrick Jensen, *A Language Older Than Words* (New York: Context Books, 2000), 126–27.

2. Bruce Lipton, *The Biology of Belief* (Santa Rosa, CA: Mountain of Love/Elite Books, 2005), 27.

3. Deepak Chopra, *The Book of Secrets* (New York: Harmony Books, 2004).

4. Paraphrased from Christopher Uhl and Dana Stuchul, *Teaching as if Life Matters: The Promise of a New Education Culture* (Baltimore: Johns Hopkins University Press, 2011).

5. Much of the following information on monarch butterfly natural history comes from Eric Grace, *The World of the Monarch Butterfly* (San Francisco: Sierra Club Books, 1997).

6. Judith Kohl and Herbert Kohl, *The View from the Oak* (New York: New York Press, 1997).

7. Robert Pyle, "Las Monarcas: Butterflies on Thin Ice," *Orion*, Spring 2001, 21.

8. The information on goldenrod ecology comes mainly from Arthur Weis and Warren Abrahamson, "Just Lookin' for a Home," *Natural History* 107 (1998): 60–63.

9. I. Bossema, "Jays and Oaks: An Eco-ethological Study of a Symbiosis," *Behavior* 70 (1979): 1–117; C. Johnson and C. S. Adkisson, "Airlifting the Oaks," *Natural History*, October 1986, 41–47.

10. Lynn Margulis, *Symbiotic Planet* (New York: Basic Books, 1998), 122.

11. Margulis, *Symbiotic Planet*, 10.

12. The Lewis Thomas quote is from Jeremy Hayward, *Letters to Vanessa* (Boston: Shambhala, 1997), 102–3.

13. Chet Raymo, *Natural Prayers* (St. Paul, MN: Hungry Mind Press, 1999).

14. See http://en.wikipedia.org/wiki/Demodex.

15. Bonnie Bassler, "How Bacteria 'Talk,'" *TED Talks*, www.ted.com/talks/bonnie_bassler_on_how_bacteria_communicate.html.

16. Daily's work is described in Andrew Beattie and Paul Ehrlich, *Wild Solutions* (New Haven, CT: Yale University Press, 2001).

17. Beattie and Ehrlich, *Wild Solutions*.

18. Thomas Berry, *The Great Work* (New York: Bell Tower, 1999), 4.

19. Lynne McTaggart, *The Intention Experiment* (New York: Free Press, 2007).

20. Rupert Sheldrake, *The Sense of Being Stared At* (New York: Crown Publishers, 2003), 4.

21. Russell Targ and J. J. Hurtak, *The End of Suffering* (Charlottesville, VA: Hampton Roads, 2006).

22. Jeremy Hayward, *Letters to Vanessa* (Boston: Shambhala, 1997), 143–44.

23. Larry Dossey, *Healing Beyond the Body: Medicine and the Infinite Reach of the Mind* (Boston: Shambhala, 2003). (Emphasis added.)

24. Marc Ian Barasch, *Field Notes on the Compassionate Life: A Search for the Soul of Kindness* (Emmaus, PA: Rodale, 2005).

25. Arthur Koestler, *The Ghost in the Machine* (New York: Penguin Books, 1990).

26. The preceding three paragraphs were paraphrased from Uhl and Stuchul, *Teaching as if Life Matters*, 165. Anthony Weston, *Back to Earth* (Philadelphia: Temple University Press, 1994).

27. Inspired in part by Lucia Capacchione, *The Art of Emotional Healing* (Boston: Shambhala, 2006).

28. Brian Swimme, *The Hidden Heart of the Cosmos* (Maryknoll, NY: Orbis Books, 1996), 56.

29. Scott Russell Saunders quoted in Evelyn Adams, "Learning to Be at Home on Fidalgo," *Orion Afield*, Autumn 2001, 10–13.

30. Stephen Talbott, "Why Is the Moon Getting Farther Away?" *Orion*, Spring 1998, 44.

PART II: ASSESSING THE HEALTH OF EARTH

1. "World Scientists' Warning to Humanity" (Boston: Union of Concerned Scientists, 1992), www.ucsusa.org.

CHAPTER 4: LISTENING

1. Thomas Berry, *The Great Work* (New York: Bell Tower, 1999), 200.

2. John Terborgh, "Why American Songbirds Are Vanishing," *Scientific American*, May 1992, 98–104.

3. Presently this survey is administered by the U.S. Geological Survey in conjunction with the Canadian Wildlife Survey.

4. Terborgh, "Why American Songbirds Are Vanishing." Also: www.mbr-pwrc .usgs.gov/bbs/bbs2009.html.

5. Alan Weisman, *The World Without Us* (New York: Thomas Dunne Books, 2007), http://birdchaser.blogspot.com/2008/03/cats-kill-over-1-billion-birds-each .html.

6. David Wilcove, "Nest Predation in Forest Tracts and the Decline of Migratory Songbirds," *Ecology* 66: 1211–14. Terborgh, "Why American Songbirds Are Vanishing."

7. Mark J. Walters, "What Bugs Us," *Orion*, January/February 2003, 14–15.

8. Alan Weisman, *The World Without Us*.

9. Christian Both, Sandra Bouwhuis, C. M Lessells, and Marcel Visser, "Climate Change and Population Declines in a Long-Distance Migratory Bird," *Nature*, May 2006, 81–83.

10. Stuart Pimm, *The World According to Pimm: A Scientist Audits the Earth* (New York: McGraw-Hill, 2001).

11. Nicky Chambers, Craig Simmons, and Mathis Wackernagel, *Sharing Nature's Interest* (London: Earthscan, 2000), 44.

12. Pimm, *The World According to Pimm*.

13. I have borrowed this analogy from Paul Ehrlich and Ann Ehrlich, *Extinction* (New York: Random House, 1981).

14. Aldo Leopold, *A Sand County Almanac* (Oxford: Oxford University Press, 1949).

15. Sylvia Earle, *The World Is Blue* (Washington, DC: National Geographic Society, 2009).

16. Tobin Hart, *The Secret Spiritual World of Children* (Novato, CA: New World Library, 2003), 47.

17. Earle, *The World Is Blue*, 17.

18. Deborah Cramer, *Smithsonian Ocean: Our Water Our World* (Washington, DC: Smithsonian Institution with HarperCollins, 2008), 114.

19. Earle, *The World Is Blue*, 15.

20. Cramer, *Smithsonian Ocean*.

21. Darlene Crist, Gail Scowcroft, and James M. Harding, *World Ocean Census* (Buffalo, NY: Firefly Books, 2009).

22. This idea of a journey from ocean surface to ocean depths was inspired by Earle, *The World Is Blue*.

23. Cramer, *Smithsonian Ocean*.

24. Ransom Myers and Boris Worm, "Rapid Worldwide Depletion of Predatory Fish Communities," *Nature*, May 15, 2003, 280–83.

25. Earle, *The World Is Blue*.

26. Tierney Thys, "For the Love of Fishes," in *Oceans: The Threats to Our Seas and What You Can Do to Turn the Tide*, ed. Jon Bowermaster (Philadelphia: Perseus Book Group, 2010), 140.

27. Wilma Subra, "Reasons to Worry about the Dead Zones," in Bowermaster, *Oceans*, 131.

28. Earle, *The World Is Blue*.

29. Earle, *The World Is Blue*.

30. David de Rothschild, "Message in a Bottle: Adventures Aboard the Plastiki," in Bowermaster, *Oceans*, 83.

31. Susan Casey, "Plastic Ocean: Our Oceans Are Turning into Plastic . . . Are We?" in Bowermaster, *Oceans*, 71.

32. Earle, *The World Is Blue*, 112.

33. Earle, *The World Is Blue*.

34. Ingrid Bacci, *The Art of Effortless Living: Simple Techniques for Healing Mind, Body and Spirit* (Croton-on-Hudson, NY: Vision Works, 2000), 110.

35. Joanna Macy and Molly Brown, *Coming Back to Life: Practices to Reconnect Our Lives, Our World* (Gabriola Island, BC: New Society Publishers, 1998).

36. Macy and Brown, *Coming Back to Life*, 27.

37. Macy and Brown, *Coming Back to Life*.

38. Bacci, *The Art of Effortless Living*, 116.

39. Paraphrased from Christopher Uhl and Dana Stuchul, *Teaching as if Life Matters: The Promise of a New Education Culture* (Baltimore: Johns Hopkins University Press, 2011).

40. Diarmuid O'Murchu, *Our World in Transition* (New York: Crossroad, 1992), 152.

41. For a fuller elaboration, see the discussion of speciesism in chapter 3.

42. This figure was inspired by Michael McKinney and Robert Schoch, *Environmental Science: Systems and Solutions* (New York: West, 1996), 596–97.

43. Barbara Smuts, "Encounters with Wild Minds," *Journal of Consciousness Studies* 8, nos. 5–7 (2001): 293–309.

44. Smuts, "Encounters with Wild Minds," 301.

45. Smuts, "Encounters with Wild Minds."

46. Marshall Rosenberg, *Nonviolent Communication: A Language of Compassion* (Encinitas, CA: PuddleDancer Press, 1999), 147.

47. See Non-Judgment Day Project (www.nonjudgmentday.org/index.html).

CHAPTER 5: COURAGE

1. Wes Nisker, *The Essential Crazy Wisdom* (Berkeley, CA: Ten Speed Press, 2001), 13.

2. Scripps CO2 Program, "Mauna Loa Record," http://scrippsco2.ucsd.edu/graphics_gallery/mauna_loa_record/mauna_loa_record.html.

3. Tyler Volk, *Gaia's Body* (New York: Copernicus/Springer-Verlag, 1998).

4. Bill McKibben, *Eaarth: Making a Life on a Tough New Planet* (New York: Henry Holt, 2010), 27–28.

5. Bill McKibben, *The End of Nature* (New York: Anchor Books, 1989), 18.

6. Just as percent means *out of a hundred*, parts per million (ppm) means *out of a million*.

7. Intergovernmental Panel on Climate Change, www.ipcc.ch/.

8. Source: NASA Goddard Institute for Space Studies, "GISS Surface Temperature Analysis," http://data.giss.nasa.gov/gistemp/graphs_v3/Fig.A.gif.

9. Intergovernmental Panel on Climate Change.

10. Greg Lankenau provided me with this *stereo* analogy, as well as the *beach* phrasing.

11. Sylvia Earle, *The World Is Blue: How Our Fate and the Ocean's Are One* (Washington, DC: National Geographic Society, 2009).

12. In a related vein, it is worth noting that a 2012 study reported that the enormous ice sheets of Greenland and Antarctica had increased their rate of melting threefold between 1990 and 2010 (www.jpl.nasa.gov/news/news.php?release=2012 -376&cid=release_2012-376).

13. Pavel Y. Groisman, Richard W. Knight, and Thomas R. Karl, "Heavy Precipitation and High Streamflow in the Contiguous United States: Trends in the Twentieth Century," *Bulletin of the American Meteorological Society* 82, no. 2 (2001): 219.

14. Michelle C. Mack, M. Syndonia Bret-Harte, Teresa N. Hollingsworth, Randi R. Jandt, Edward A. G. Schuur, Gaius R. Shaver, and David L. Verbyla, "Carbon Loss from an Unprecedented Arctic Tundra Wildfire," *Nature* 475 (2011): 489–92.

15. Joe Romm, "The Permafrost Won't Be Perma for Long," *Think Progress*, May 22, 2008, http://thinkprogress.org/romm/2008/05/22/202419/tundra-part -1-the-permafrost-wont-be-perma-for-long/ and Joe Romm, "For Peat's Sake: A Point of No Return as Alarming as the Tundra Feedback," ClimateProgress.org, October 1, 2008.

16. Tom Knudson, "Sierra Nevada Climate Changes Feed Monster, Forest-Devouring Fires," *Sacramento Bee*, November 30, 2008.

17. Jim Robbins, "Bark Beetles Kill Millions of Acres of Trees in the West," *New York Times*, November 18, 2008.

18. Matthew Daly, "House Approves Special Spending to Fight Wildfires," Associated Press, March 26, 2009.

19. David Adam, "Amazon Could Shrink by 85% Due to Climate Change, Scientists Say," *Guardian,* www.guardian.co.uk/environment/2009/mar/11/amazon -global-warming-trees.

20. Deborah Cramer, *Smithsonian Ocean: Our Water Our World* (New York: Smithsonian in association with HarperCollins, 2008).

21. Lisa Suatoni, "Acid Test," in *Oceans: The Threats to Our Seas and What You Can Do to Turn the Tide*, ed. Jon Bowermaster (Philadelphia: Perseus Book Group, 2010), 103–6.

22. Earle, *The World Is Blue*, 179.

23. McKibben, *Eaarth*, xiii.

24. Water Encyclopedia, "Ogallala Aquifer," www.waterencyclopedia.com/Oc-Po/Ogallala-Aquifer.html.

25. Soil erosion data are for the U.S. cropland average for nonfederal rural land, accessed at www.nrcs.usda.gov.

26. David Pimentel, "Soil Erosion: A Food and Environmental Threat," *Journal of the Environment, Development and Sustainability* 8 (2006): 119–37.

27. Pimentel, "Soil Erosion."

28. Theo Colburn, "Toxic Legacy," *Yes!* Summer 1998, 14–18.

29. Sandra Steingraber, Penn State University Public Lecture, October 23, 2000.

30. Steingraber, Penn State University Public Lecture.

31. Sandra Steingraber, "PVC and the Breasts of Mothers," *Rachel's Environment and Health*, July 1999, www.rachel.org/?q=en/node/4833.

32. Sandra Steingraber, *Living Downstream: A Scientist's Personal Investigation of Cancer and the Environment* (New York: Vintage Books, 1998).

33. This figure was inspired by Theo Colburn, Diane Dumanoski, and J. Peter Myers, *Our Stolen Future* (New York: Dutton, 1996), 72.

34. Colburn et al., *Our Stolen Future*.

35. Lifesite News, "Study Confirms Estrogen in Water from the Pill Devastating to Fish Population," February 18, 2008, www.lifesitenews.com/news/archive/ldn/2008/feb/08021805.

36. Laura Vandenberg, Theo Colburn, Tyrone Hayes, Jerrold Heindel, David Jacobs Jr., Duk-Hee Lee, Toshi Shioda, Ana Soto, Frederick vom Saal, Wade Welshons, Thomas Zoeller, and John Myers, "Hormones and Endocrine-Disrupting Chemicals: Low-Dose Effects and Nonmonotonic Dose Responses," *Endocrine Review* 33 (2012): 378–455.

37. Nicholas D. Kristof, "It's Time to Learn from Frogs," *New York Times*, June 27, 2009, www.nytimes.com/2009/06/28/opinion/28kristof.html.

38. Michael Zimmerman, *Science, Nonscience, and Nonsense: Approaching Environmental Literacy* (Baltimore: Johns Hopkins University Press, 1995), 117–18.

39. Derrick Jensen, *A Language Older Than Words* (New York: Context Books, 2000), 2.

40. My interest in and concern for honeybees has prompted me to become an urban beekeeper. I now have two hives in my backyard. I had a smidgen of book-learning when I began, but nothing compared to the delights I have experienced by living in close proximity with bees, watching them visit the flowers in our family vegetable garden; observing them as they return to the hive, their leg sacks filled with pollen; marveling at their capacity to cooperate for the good of the whole. There have been moments of sadness too, such as when both my colonies collapsed not long ago.

41. Wendell Berry, *A Timbered Choir* (Boulder, CO: Counterpoint Press, 1998).

42. Dictionary.com, http://dictionary.reference.com/browse/precautionary+principle.

43. David Bohm, Donald Factor, and Peter Garrett, "Dialogue: A Proposal," www.david-bohm.net/dialogue/dialogue_proposal.html.

44. Tamarack Song, *Sacred Speech: The Way of Truthspeaking* (Three Lakes, WI: Teaching Drum Outdoor School, 2004), i.

45. Song, *Sacred Speech.*

46. Song, *Sacred Speech*, 69, 84, 85.

47. Inspired by Austin Mandryk.

48. Peter Block, *Community and the Structure of Belonging* (San Francisco: Berrett-Koehler, 2008).

CHAPTER 6: LIVING THE QUESTIONS

1. Neil Postman and Charles Weingartner, *Teaching as a Subversive Activity* (New York: Delta, 1969), 23.

2. Fritz Stern, "The Importance of 'Why,'" *World Policy Journal*, Spring 2000, 1–8.

3. Bill McKibben, *Deep Economy: The Wealth of Communities and the Durable Future* (New York: Henry Holt, 2007); "Kerala: Human Development Report," State Planning board, Government of Kerala, 2006; http://en.wikipedia.org/wiki/Kerala#Human_Development_Index.

4. Bill McKibben, "The Enigma of Kerala," *Utne Reader*, March–April 1996, 103–11.

5. McKibben, "The Enigma of Kerala."

6. Wikipedia, "Kerala," http://en.wikipedia.org/wiki/Kerala#Human_Development_Index; McKibben, *Deep Economy*.

7. Ellen J. Langer, *Mindfulness* (Cambridge, MA: Perseus Books, 1989), 9.

8. Sources for graphs include:
   Population: http://geography.about.com/od/obtainpopulationdata/a/world population.htm
   Gross world product: http://en.wikipedia.org/wiki/Gross_world_product
   Water: www.chess.com/forum/view/off-topic/water--the-new-oil?page=4
   Paper: www.ecology.com/2011/09/10/paper-chase/
   Motor vehicles: http://en.wikipedia.org/wiki/Motor_vehicle
   Mobile phones: http://en.wikipedia.org/wiki/Mobile_phone

9. Bill McKibben, "A Special Moment in History," *Atlantic Monthly*, May 1998, 55–78.

10. Eugene Odum, *Ecological Vignettes* (Amsterdam: Harwood, 1998).

11. Juliet Schor, *Plentitude* (New York: Penguin Press, 2010).

12. Schor, *Plentitude.*

13. James Speth, *The Bridge at the Edge of the World: Capitalism, the Environment, and Crossing from Crisis to Sustainability* (New Haven, CT: Yale University Press, 2008); Schor, *Plentitude.*

14. Schor, *Plentitude*.

15. Richard Layard, *Happiness: Lessons from a New Science* (New York: Penguin Press, 2005).

16. Pacific Institute, "Bottled Water and Energy: A Fact Sheet," www.pacinst.org/topics/water_and_sustainability/bottled_water/bottled_water_and_energy.html.

17. Emily Arnold and Janet Larsen, "Bottled Water: Pouring Resources Down the Drain," Earth Policy Institute, February 2, 2006, www.earth-policy.org/index.php?/plan_b_updates/2006/update51.

18. National Resources Defense Council, *Bottled Water: Pure Drink or Pure Hype?* www.nrdc.org/water/drinking/bw/exesum.asp.

19. David de Rothschild, "Message in a Bottle: Adventures Aboard the Plastiki," in *Oceans: The Threats to Our Seas and What You Can Do to Turn the Tide*, ed. Jon Bowermaster (New York: PublicAffairs, 2010), 83–91.

20. Andrew Theen, "Ivy Colleges Shunning Bottled Water Jab at $22 Billion Industry" March 7, 2012, Bloomberg.com, www.bloomberg.com/news/2012-03-07/ivy-colleges-shunning-bottled-water-jab-at-22-billion-industry.html.

21. Energy constitutes the lion's share of most people's footprint. Analysts calculate the energy footprint of a product by first determining the amount of fossil fuel needed to make the product along with the quantity of CO_2 released in the burning of that fossil fuel. Then, they calculate the land area required to absorb the said amount of CO_2. This approach is taken because the unrestricted release of carbon dioxide into the atmosphere has begun to destabilize Earth's climate (see chapter 5). Hence, to stabilize global climate, land must be set aside to assimilate the carbon dioxide released in the burning of fossil fuels.

22. Wikipedia, "List of Countries by Ecological Footprint," http://en.wikipedia.org/wiki/List_of_countries_by_ecological_footprint.

23. Global Footprint Network, *Global Footprint Network 2007 Annual Report*, www.footprintnetwork.org.

24. Global Footprint Network, *Global Footprint Network 2007 Annual Report*.

25. Global Footprint Network, "Overshoot Trends," www.footprintnetwork.org/en/index.php/GFN/page/overshoot_trends.

26. The *Limits to Growth* study was revised in 2004 with generally similar findings (Donella Meadows, Jorgen Randers, and Dennis Meadows, *Limits to Growth: The 30-Year Update* (White River Junction, VT: Chelsea Green Publishers, 2004).

27. Jared Diamond, *Collapse* (New York: Viking, 2005), 504.

28. Source unknown.

29. Lester Brown, *Plan B 3.0: Mobilizing to Save Civilization* (New York: W. W. Norton, 2008); Index Mundi, "World Crude Oil Consumption by Year," www.indexmundi.com/energy.aspx; Janet Larson, "Bumper 2011 Grain Harvest Fails to Rebuild Global Stocks," *Eco-Economy Indicators*, January 11, 2012, Earth Policy Institute, www.earth-policy.org/indicators/C54/grain_2012.

30. Happy Planet Index, www.happyplanetindex.org/.

31. This framing was suggested to me by Gregory Lankenau.

32. Wendell Berry, *The Long-Legged House* (New York: Harcourt, Brace & World, 1969).

33. Martha Beck, *Finding Your Own North Star: Claiming the Life You Were Meant to Live* (New York: Three Rivers, 2001), 162.

34. Rainer Maria Rilke, *Letters to a Young Poet*, translated by M. D. Herter Norton (New York: W. W. Norton, 1993), 33, 35.

35. Jerry Mander, *In the Absence of the Sacred: The Failure of Technology and the Survival of the Indian Nations* (San Francisco: Sierra Club Books, 1991), 43–44.

36. Derrick Jensen, *A Language Older Than Words* (New York: Context Books, 2000), 40.

37. Chellis Glendinning, *My Name Is Chellis and I'm in Recovery from Western Civilization* (Boston: Shambhala, 1994).

PART III: HEALING OURSELVES, HEALING EARTH

1. Jelaluddin Rumi, in *Open Secret*, translated by John Moyne and Coleman Barks (Boston: Shambhala), 7.

2. Eckhart Tolle, *The Power of Now* (Novato, CA: New World Library, 1999), 67.

CHAPTER 7: THE OLD STORY

1. Quoted in Sabrina Hassumani, *Salman Rushdie: A Postmodern Reading of His Major Works* (Cranbury, NJ: Associated University Presses, 2002), 104.

2. Charles Eisenstein, *The Ascent of Humanity* (Harrisburg, PA: Panenthea Press, 2007); Joseph Chilton Pearce, *The Biology of Transcendence: A Blueprint of the Human Spirit* (Rochester, VT: Park Street Press, 2002); Bill Plotkin, *Nature and the Human Soul: Cultivating Wholeness and Community in a Fragmented World* (Novato, CA: New World Library, 2008).

3. Daniel Quinn, *Ishmael* (New York: Bantam, Turner, 1992).

4. Quinn, *Ishmael*, 35–36.

5. Barbara M. Hubbard, *Conscious Evolution: Awakening the Power of Our Social Potential* (Novato, CA: New World Library, 1998), 178.

6. Quinn, *Ishmael*.

7. Quinn, *Ishmael*.

8. Paraphrased from Christopher Uhl and Dana Stuchul, *Teaching as if Life Matters: The Promise of a New Education Culture* (Baltimore: Johns Hopkins University Press, 2011).

9. Big Ten Science, "Chemical Found in Sanitizer May Weaken Immune System," January 6, 2011, http://bigkingken.wordpress.com/2011/01/06/chemical-found-in-sanitizer-may-weaken-immune-system.

10. Uhl and Stuchul, *Teaching as if Life Matters*.

11. Barbara Brandt, *Whole Life Economics* (Gabriola Island, BC: New Society Publishers, 1995).

12. This story is paraphrased from Susan Strauss, *The Passionate Fact* (Golden, CO: North American Press, 1996), 9.

13. Brandt, *Whole Life Economics*.

14. Jerry Mander, *In the Absence of the Sacred: The Failure of Technology and the Survival of the Indian Nations* (San Francisco: Sierra Club Books, 1991).

15. Denise Pope, *Doing School: How We Are Creating a Generation of Stressed Out, Materialistic, and Miseducated Students* (New Haven, CT: Yale University Press, 2001).

16. Harwood Group, "Yearning for Balance," *Yes!* Spring/Summer 1996, 17.

17. Charles Eisenstein, *Sacred Economics: Money, Gift and Society in the Age of Transition* (Berkeley, CA: Evolver Editions, 2011).

18. Note: 1 gigaton is equivalent to one billion metric tons. This number, 565 gigatons, isn't exact, but as climate models continue to be refined, year-by-year, the adjustments, up or down, are increasingly small.

19. Bill McKibben, "Global Warming's Terrifying New Math," *Rolling Stone*, July, 19, 2012, www.rollingstone.com/politics/news/global-warmings-terrifying-new-math-20120719.

20. Carbon Tracker Initiative, www.carbontracker.org.

21. McKibben, "Global Warming's Terrifying New Math."

22. McKibben, "Global Warming's Terrifying New Math."

23. For more information on the potential deleterious effects of hydrofracking, see Dina Cappiello, "Fracking Impacts Could Be 'Excessive,' Federal Advisory Panel Warns," *Huffington Post*, November 10, 2011, www.huffingtonpost.com/2011/11/10/panel-says-fracking-impacts-excessive_n_1087049.html.

24. Eisenstein, *Sacred Economics*.

25. Bill McKibben, *Deep Economy: The Wealth of Communities and the Durable Future* (New York: Henry Holt, 2007); Robert Putnam, *Bowling Alone* (New York: Simon & Schuster, 2000).

26. Juliet Schor, *Plentitude* (New York: Penguin Press, 2010).

27. Benjamin Hunnicutt, "When We Had the Time," in *Take Back Your Time*, ed. John de Graaf (San Francisco: Berrett-Koehler Publishers, 2003), 115.

28. Quoted in de Graaf, *Take Back Your Time*.

29. McKibben, *Deep Economy*.

30. McKibben, *Deep Economy*, 108.

31. Eisenstein, *Sacred Economics*.

32. Eisenstein, *Sacred Economics*.

33. Frances Moore Lappe, Joseph Collins, and Peter Rosset, *World Hunger: Twelve Myths* (New York: Grove Press, 1998).

34. Eisenstein, *Sacred Economics*.

35. Marc Ian Barasch, *Field Notes on the Compassionate Life: A Search for the Soul of Kindness* (Emmaus, PA: Rodale, 2005).

36. McKibben, *Deep Economy*.

37. McKibben, *Deep Economy*.

38. Ed Diener and Martin Seligman, "Beyond Money: Toward an Economy of Well-Being," *Psychological Science in the Public Interest* 5, no. 1 (2004): 1–31.

39. David Myers, "What Is the Good Life?" *Yes!* Summer 2004, 15.

40. Don Miguel Ruiz, *The Four Agreements: A Practical Guide to Personal Freedom* (San Rafael, CA: Amber-Allen Publishing, 1997).

41. Charlene Spretnak, *Resurgence of the Real* (New York: Routledge, 1999), 56.

42. Byron Katie, *Loving What Is* (New York: Three Rivers Press, 2002), 6–7.

43. Berry, *The Dream of the Earth*, 171.

44. Ken Wilbur, *A Brief History of Everything* (Boston: Shambhala, 1996), 69.

45. Wilbur, *A Brief History of Everything*.

46. Katie, *Loving What Is*.

47. Peter Senge, C. Otto Scharmer, Joseph Jaworski, and Betty Flowers, *Presence: Human Purpose and the Field of the Future* (Cambridge, MA: Society for Organizational Learning, 2004), 27.

48. Inspired by "What-the-Bleep Study Guide," www.whatthebleep.com/guide.

49. Gregg Levoy, *Callings: Finding and Following an Authentic Life* (New York: Three Rivers Press, 1997), 139.

CHAPTER 8: BIRTHING A NEW STORY

1. Ursula K. Le Guin, *Voices* (New York: Harcourt Brace, 2006), 166–67.

2. Katrina Shields, *In the Tiger's Mouth* (Gabriola Island, BC: New Society, 1994), 12.

3. Michael Dowd, *Thank God for Evolution: How the Marriage of Science and Religion Will Transform Your Life and Our World* (New York: Viking, 2008), 1.

4. David Abram *Becoming Animal: An Earthly Cosmology* (New York: Pantheon, 2010).

5. Anthony Weston, *Back to Earth* (Philadelphia: Temple University Press, 1994).

6. Abram, *Becoming Animal*.

7. Richard Nelson, *The Island Within* (New York: Vintage, 1991), 249.

8. Joanna Lauck, *The Voice of the Infinite in the Small: Re-Visioning the Insect-Human Connection* (Boston: Shambhala, 2002), 42.

9. Sharif Abdullah, *Creating a World That Works for All* (San Francisco: Berrett-Koehler, 1999), 80.

10. Daniel Goleman, *Social Intelligence* (New York: Bantam, 2006).

11. Goleman, *Social Intelligence*, 43.

12. Marc Barasch, *Field Notes on the Compassionate Life: A Search for the Soul of Kindness* (Emmaus, PA: Rodale, 2005).

13. Leslie Brothers, "A Biological Perspective on Empathy," *American Journal of Psychiatry* 146, no. 1 (1989): 10–19.

14. Michael Tomasello, *Why We Cooperate* (Cambridge, MA: MIT Press, 2009).

15. Charles Eisenstein, *Sacred Economics: Money, Gift and Society in the Age of Transition* (Berkeley, CA: Evolver Editions, 2011).

16. Angie O'Gorman, "Defense through Disarmament: Nonviolence and Personal Assault," in *The Universe Bends Toward Justice*, ed. Angie O'Gorman (Gabriola Island, BC: New Society, 1990), 242–46.

17. Richard Rohr, *New Great Themes of Scripture* (Cincinnati, OH: St. Anthony Messenger Press, 1999).

18. Abdullah, *Creating a World That Works for All*.

19. Previous three paragraphs paraphrased from Uhl and Stuchul, *Teaching as if Life Matters*.

20. Anthony DeMello, *Awareness: The Perils and Opportunities of Reality* (New York: Image Books, 1992).

21. Gus Speth, *The Bridge at the Edge of the World: Capitalism, the Environment, and Crossing from Crisis to Sustainability* (New Haven, CT: Yale University Press, 2008), 213–14.

22. Charles Eisenstein, *The Ascent of Humanity* (Harrisburg, PA: Panenthea Press, 2007); Thomas Berry, *The Great Work* (New York: Bell Tower, 1999); Bill Plotkin, *Nature and the Human Soul: Cultivating Wholeness and Community in a Fragmented World* (Novato, CA: New World Library, 2008).

23. David Korten, *The Great Turning: From Empire to Earth Community* (Bloomfield, CT: Kumarian Press, 2007).

24. Shannon Hayes, *Radical Homemakers: Reclaiming Domesticity from a Consumer Culture* (Richmondville, NY: Left to Write Press, 2010), 13.

25. John De Graff (ed.), *Take Back Your Time* (San Francisco: Berrett-Koehler Publishers, 2003).

26. Robert Putnam, *Bowling Alone: The Collapse and Revival of American Community* (New York: Simon and Schuster, 2010).

27. Juliet Schor, *Plentitude* (New York: Penguin Press, 2010).

28. Erin Middlewood, "Social Medicine," *Orion*, September–October 2005, 26, reported in Bill McKibben, *Deep Economy: The Wealth of Communities and the Durable Future* (New York: Henry Holt, 2007).

29. From a strictly economic perspective it can be argued that time is money insofar as electing to spend one's time outside of the wage economy results in a loss of income—that is, there is an opportunity cost for withdrawing from the cash economy. But from the perspective of a person choosing to withdraw from the economy, this so-called opportunity cost has little, if any, relevance.

30. Eisenstein, *Sacred Economics*.

31. Schor, *Plentitude*.

32. International Downshifting Week, "America: A Little Downshifting Never Hurt Anyone," March 29, 2011, http://downshiftingweek.wordpress.com/2011/03/29/america-a-little-downshifting-never-hurt-anyone.

33. Schor, *Plentitude*, 101–2.

34. Schor, *Plentitude*.

35. McKibben, *Deep Economy*.

36. To locate CSAs and farmer's markets in your area go to www.localharvest.org.

37. Rachel Botsman and Roo Rogers, *What's Mine Is Yours: The Rise of Collaborative Consumption* (New York: HarperCollins Publishers, 2010).

38. Brian Halweil, *Eat Here: Reclaiming Homegrown Pleasures in a Global Supermarket*, (Washington, DC: Worldwatch Institute, 2004).

39. Bill McKibben, *Eaarth* (New York: Henry Holt, 2010).

40. Greg Pahl, *Citizen-Powered Energy Handbook* (White River Junction, VT: Chelsea Green, 2007).

41. Laura Sevier, Mike Henderson, and Nritijuna Naidu, "Ecovillages: A Model Life?" *Ecologist* 38, no. 4 (2008): 36–41.

42. See Transition United States at http://transitionus.org/initiatives-map.

43. Transition United States, "Transition 101," http://transitionus.org/transition-101.

44. Eisenstein, *Sacred Economics*, 78.

45. Botsman and Rogers, *What's Mine Is Yours*.

46. Botsman and Rogers, *What's Mine Is Yours*.

47. Botsman and Rogers, *What's Mine Is Yours*, 163.

48. Jim Salter, "Pay-What-You-Want Panera Called a Success," *USA Today*, May 16, 2011, www.usatoday.com/money/industries/food/2011-05-16-panera-pay-what-you-can_n.htm.

49. Botsman and Rogers, *What's Mine Is Yours*.

50. Paul Hawken, Amory Lovins, and Hunter Lovins, *Natural Capitalism: Creating the Next Industrial Revolution* (Boston: Little, Brown, 1999).

51. Botsman and Rogers, *What's Mine Is Yours*.

52. Quoted at *Science and Philosophy*, http://sciphilos.info/docs_pages/docs_Einstein_fulltext_css.html.

53. Marshall Rosenberg, *Nonviolent Communication: A Language of Compassion* (Encinitas, CA: PuddleDancer Press, 2001).

54. Eisenstein, *Sacred Economics*, 3–4.

55. Margaret Wheatley and Deborah Frieze, *Walk Out Walk On* (San Francisco: Berrett-Koehler Publishers, 2011), 140.

56. Eisenstein, *Sacred Economics*.

57. Eisenstein, *Sacred Economics*.

58. Eisenstein, *Sacred Economics*, 358.

59. Bill Plotkin, *Soulcrafting* (Novato, CA: New World Library, 2003), 59.

60. Puanani Burgess, "Blessings Revealed" *Yes!* Winter 2009.

61. Paraphrased from Eisenstein, *Sacred Economics*, 358.

62. Wendell Berry, *Conversations with Wendell Berry*, ed. Morris Allen Grubbs (Jackson: University Press of Mississippi, 2007).

63. Margaret Wheatley, *Leadership and the New Science* (San Francisco: Berrett-Koehler, 1999).

64. Wheatley, *Leadership and the New Science*, 119.

65. Tara Brach, *Radical Acceptance: Embracing Your Life with the Heart of the Buddha* (New York: Bantam Books, 2003), 25.

66. Mark A. Burch, *Stepping Lightly* (Gabriola Island, BC: New Society, 2000), 106.

67. Wes Nisker, *Buddha's Nature* (New York: Bantam Books, 1998), 26.

68. Nisker, *Buddha's Nature*, 26.

69. Wheatley and Frieze, *Walk Out Walk On*, 157.

70. Eisenstein, *Sacred Economics*.

71. For ideas, go to: http://toponlineengineeringdegree.com/?page_id=116.

72. Project for Public Spaces, www.pps.org/about/team/jwalljasper.

73. Wheatley and Frieze, *Walk Out Walk On*, 58.

EPILOGUE

1. Clarissa Pinkola Estes, "Inspiration," Aloha International, www.huna.org/html/cpestes.html.

2. Elisabet Sahtouris, "The Butterfly Story," *Lifeweb*, www.sahtouris.com.

3. Susan Griffin, "Can Imagination Save Us?" *Utne Reader*, July–August 1996, 43–45.

4. Charles Eisenstein, *Sacred Economics: Money, Gift and Society in the Age of Transition* (Berkeley, CA: Evolver Editions, 2011), 445.

5. Eisenstein, *Sacred Economics*.

6. Inspired by Eisenstein, *Sacred Economics*.

INDEX

Abram, David, 32–33, 207
abundance, versus economism, 193–94
acceptance, 170; cells and, 56
ADHD, nature deficit and, 49
adolescence, of culture, 239–40
airscape, 118–20
algae, 70
Amazon, 127
American Paradox, 195
anglerfish, 101
aphids, 65–66, 67*f*
applications. *See* practices
Arctic, 124–25
Aristotle, 181–82
assumptions, questioning, 167–69
awakening: by doing, 233–35;
    mindfulness and, 231–32
awareness, 29–53; cells and, 56;
    development of, 109, 110*f*;
    expanding, 49–53; and pain, 105–7

baboons, 111
Bacci, Ingrid, 105, 107
bacteria, 69, 71–72; surface area of, 35

Barasch, Marc, 210
bees, 137–39, 251n40
beetles, 46, 126
Berry, Thomas: on community, 75; on
    listening, 89; on story, 22, 47, 198;
    on universe, 5
Berry, Wendell, 139, 165, 229
Big Bang, 9–11; causes of, 11–12
big picture: awareness and, 47–49;
    community and, 75–78; courage
    and, 137–39; discovery and, 19–22;
    economism and, 196–98; listening
    and, 105–7; new story and, 225–29;
    questions and, 162–64
Biosphere 2, 74
birds, 68–69, 90–96
birthing: new culture, 237–41; new
    story, 205–36
blue jays, 68–69
Bohm, David, 140
Bormann, Herbert, 43
bottled water, 157–58
Brach, Tara, 230
Braud, William, 76–77

breastfeeding, 132
Burch, Marc, 231
Burgess, Puanani, 228
butterflies, 58–64, 237–38
bycatch, 102–3

cancer, 57, 131–33
carbon dioxide, atmospheric, 118–20,
    119f, 121–26, 189
cats, 92–93
cells: characteristics of, 56–57; origins
    of, 16
China, 163
chlorophyll, 36
Chopra, Deepak, 19, 56–57
chronos, 26
clay, 81–82
climate change, 117–45; denial of, 122;
    effects of, 124–29; and migratory
    birds, 94; research on, 121–24; state
    of, 118–29
collaborative economy, 222–25
collective consciousness, 210–11
common sense, versus economism,
    187–88
community, 55–87; Berry on, 75, 229;
    cells and, 56; economy and, 215–17;
    and food and energy needs, 218–20;
    new story on, 214–25; and shelter
    needs, 221–22; social capital and,
    216–18; thought experiment on,
    72–75
community-supported agriculture
    (CSA), 219
connection, 2–3; dissolving boundaries
    and, 211; interdependence, 206–14;
    with life, 79–80; with ocean, 98–99;
    with something greater, 5–28
consciousness: collective, 210–11. *See
    also* ecological consciousness
consumption, 153–58, 154f; consumer
    lifestyle, 153–55; questions and,
    169–71; reasons for, 155–56

coral reefs, 127–28
corn, 37–39
Cosmic Background Explorer, 10
courage, 117–45; Berry on, 139
Cowan, Stuart, 47
Craigslist, 223
Cramer, Deborah, 99
craving, 156
CSA. *See* community-supported
    agriculture
cycles: awareness and, 52–53; of Earth,
    41–47

Daily, Gretchen, 72–73
Dalai Lama, 208
dead zones, 103–4
death, 52–53
decomposers, 38–40
DeMello, Anthony, 213
denial. *See* courage
Desnos, Robert, 238
despair, 106, 118
Diamond, Jared, 161
discovery, 5–28; cultivating, 23–26
diversity, of life on Earth, 17
Dossey, Larry, 78
Dowd, Michael, 207
downshifting, 217–18
dualism, 181
Duncan, David James, 29, 50
Dunne, Brenda, 77

Earle, Sylvia, 98, 128
Earth, 24–25; assessing health of, 85–
    87, 117–18; awareness and, 51–53;
    breath of, 118–20; cycles of, 41–47;
    economism and, 190–91; healing,
    173–75; intimacy with, 80–84; life
    on, 15–17; metabolism of, 36–40;
    as mother, 2, 29–40; movement of,
    7; natural capital of, 129–30, 161;
    ocean-centric view of, 96–97, 97f;
    old story on, 179–80; productive

acreage of, 160*t*; seeing with new eyes, 207–8; surface of, 33–34, 34*f*
ecocentrism, 109
ecological consciousness, expanding: awareness and, 49–50; community and, 79–80; courage and, 139–40; discovery and, 22–23; economism and, 198–99; listening and, 108–9; new story and, 229–30; questions and, 165–66
ecological footprint, 158–61
economism, 177–203; consequences of, 187–95; nature of, 182–87; as pseudo-religion, 185–87
economy: collaborative, 222–25; gift, 32, 225–29, 235; new story on, 215–17
eco-villages, 221–22
education, economism and, 183–84
Einstein, Albert, 11, 14, 225
Eisenstein, Charles, 222, 227
emotion, 105–8, 209–10
endocrine disrupters, 133, 135–37
energy, 253n21; community and, 218–20; of Sun, 14, 36–37
Enlightenment, 196
Epictetus, 110
Escherichia coli, 71
Estes, Clarissa Pinkola, 237
exchange, 33–36
extinction, 95–96; consequences of, 96

Factor, Donald, 140
fair-earth share, 162
family size, 150, 162
farmer's markets, 219–20
fear, 106, 193–94
Fifty Questions, 81
fish, 96–105
Fisher, Charles, 102
fisheries, 102–3
flatworms, 70
fleas, 72

food, community and, 218–20
footprinting, 158–64, 253n21
forests: of Amazon, 127; of American West, 126–27; decomposers and, 38; fragmentation of, 93; nutrient cycle and, 43–45, 44*f*
fossil fuels, 120, 188–90
fracking, 190
Freecycle, 223–24
Fuller, Buckminster, xiii
fungi, 70–71

galaxies, number of, 9
gallflies, 66–68, 67*f*
Garrett, Peter, 140
geography, of home place, 82–84
giant squid, 101
gift circles, 233–34
gift economy, 32, 225–29, 235
giving, cells and, 56
Glendinning, Chellis, 48, 169
global warming, 120–21, 123*f*; research on, 121–24. *See also* climate change
God, 21–22
goldenrod, 64–68, 67*f*
Goleman, Daniel, 209–10
Goodman, Ellen, 192
gravity, 24–25, 31
Great Turning, 205–41
greed, 193–94
greenhouse gases, 120, 126, 128
Guth, Alan, 11

habituation, 156
happiness: consumption and, 164; versus economism, 194–95
Happy Planet Index (HPI), 164
Havel, Vaclav, ix
Haydn, Joseph, 25
health, 130–37
higher purpose, cells and, 57
Holdren, John, 73
holon, 78

home place, geography of, 82–84
honeybees, 137–39, 251n40
house size, 192
HPI. *See* Happy Planet Index
Hubbard, Barbara, 179
Hubbard Brook ecosystem, 43–45, 44*f*
Hubble Space Telescope, 6
humans: development of, chemicals
    and, 133–37, 134*f*; population
    issues, 148–58; and web of life,
    71–72
hydrofracking, 190
hyperindividualism, 192–93

interdependence, 206–14
Intergovernmental Panel on Climate
    Change (IPCC), 123
intimacy, with earth, 80–84

Jahn, Robert, 77
Jensen, Derrick, 55, 137, 168
judgments: listening and, 111–15;
    truthspeaking and, 141

Kabat-Zinn, Jon, 51–52
kairos, 26
Katie, Byron, 197, 199–200
Kerala, 150–52; quality-of-life
    indicators in, 151*t*
Klinenberg, Eric, 217
Koestler, Arthur, 78
Korten, David, 215
Krafel, Paul, 40
Kumar, Satish, 226

Langer, Ellen, 152
Lauck, Joanne, 209
Le Guin, Ursula K., 205
Lemley, Brad, 11
Leopold, Aldo, 95–96
less-developed countries (LDCs), 150,
    162; quality-of-life indicators in,
    151*t*

Levi, Primo, 147
Levoy, Greg, 202
life: in corn field, 37–39; exchange and,
    33–36; history of, 16–17; as lifting
    up and flowing down, 39–40; origins
    of, 15–16; on other planets, 17–19;
    relationships and, 64–75; tree of,
    79–80; web of, 57–64
Likens, Gene, 43
Linde, Andrei, 12
Lipton, Bruce, 56
listening, 89–115; Berry on, 89; to
    birds, 90–96; to fish, 96–105
living machine, 46–47
longevity, 216
Louv, Richard, 48–49
Lyme disease, 93

Macy, Joanna, 106, 209
madura foot, 72
Mander, Jerry, 167
Margulis, Lynn, 16, 69
matter, versus emptiness, 35
McKibben, Bill, 151, 192–93
meditation, 231–32
methane, 126
migration: birds and, 90–94; monarchs
    and, 61–64, 62*f*; reasons for, 91
milkweed, 60
Milky Way Galaxy, 6, 23–24, 24*f*
mindfulness, 51–52; cultivating, 231–32
mirror neurons, 210
mites, 71–72
Mohini, 230
monarch butterflies, 58–64, 237
money, 184. *See also* economism;
    economy
Moore, Charles, 104
moose, 66
mosquitoes, 32, 55–56
Muir, John, 49
mutualism, 68–71
Myers, David, 195

natural capital, 129–37, 161

nature, lack of exposure to, 47–79

needs: community and, 218–22; manufactured, 157–58; versus wants, 169–70

neighborhoods, 234–35

Nelson, Richard, 208

neonicotinoids, 138

new story, 22–23, 237–41; birthing, 205–36; on origin of universe, 8–12

Nisker, Wes, 117, 232

Noah Principle, 109

North Pacific Gyre, 104

novelty, 156

nutrient cycle, 43–45, 44*f*

oaks, 68–69

ocean(s), 97*f*; condition of, 102–5; coral reefs, 127–28; creatures of, listening to, 96–105; layers of, 99–102; surface area of, 34–35, 34*f*

Odum, Eugene, 153

O'Gorman, Angie, 211

old story, 22–23, 177–203; on origin of universe, 7–8; sustainability and, x–xi

O'Murchu, Diarmuid, 108

open-source software, 227

Other: dissolving boundaries with, 211; listening to, 110–15; seeing with new eyes, 209–12

Panera Bread Company, 224

panspermia, 15

participation: and economy, 222–25; practices for, 233–35. *See also* connection

peat, 126

peer-to-peer renting, 223

personal competency, 234

photons, 36

photosynthesis, 36, 39

phytoplankton, 34–35, 100

Pimm, Stuart, 95

pine beetle, 126

planets, formation of, 12–13

plastic, 104–5

Plotkin, Bill, 228

pollinators, 137–39

population issues, 148–58; and consumption, 153–58, 154*f*; density, 192; doubling, 148–49

Postman, Neil, 147

practices: for awareness, 50–53, 231–35; for community, 80–84; for courage, 140–45; for discovery, 23–26; for listening, 110–15; for living the questions, 166–71; for new story, 231–35; for story, 199–203

Precautionary Principle, 139–40

Prechtel, Martin, 210

Prigogine, Ilya, 229

Proust, Marcel, 1

purchasing decisions, questions and, 170–71

Pyle, Robert Michael, 61

quantum entanglement, 78

questions, 81; fresh, 152; and healing Earth, 174–75; and identity, 212–14; living, 147–71. *See also* reflection

Quinn, Daniel, 178

rainforest, 127

Raymo, Chet, 6, 25

recycling, 39

reflection, 27–28; on awareness, 53; on community, 84; on courage, 145; on economism, 203; on listening, 115; on new story, 236; on questions, 171; on universe, 26–27

relationships: Earth and, 30; importance of, 206–14; mutualism and symbiosis, 68–71; universe and, 20, 23–25; and web of life, 64–75

religion, 196; economism as, 185–87; and science, 21–22

Rilke, Rainer Maria, 166–67
Rohr, Richard, 212
Rosenberg, Marshall, 113–14
Roseto effect, 216
Ruiz, Miguel, 196
Rule-70, 149
Rumi, Jelaluddin, 173
ruminants, 69

Sagan, Carl, 12
Sahtouris, Elisabet, 21, 237
Sanders, Scott Russell, 83
Schor, Juliet, 217–18
Schweickart, Rusty, 2
science, demystifying, 58–64
scientific method, 57–64
seasons, relationship to, 25–26
secularism, 196
seeing with new eyes: Earth, 207–8;
    human Other, 209–12; ourselves,
    212–14
separation, 2–3, 177–203; age of,
    180–82
Shaich, Ron, 224
shelter, community and, 221–22
Shields, Katrina, 205
silence, 105–7
singularity, 9
Smuts, Barbara, 111
social capital, 216–17; creating, 217–18
soil, 129–30
solar system, origins and workings of,
    12–19
Song, Tamarack, 141–42
speciesism, 79
Speth, Gus, 214
spiders, 80
squid, 101
stars: distance between, 19; formation
    of, 12–13; watching, 23–25
story: Berry on, 22, 47, 177, 198; in
    daily life, 201–2; harvesting, 202;

impact of, 178–79; power of, 199–
    203. See also new story; old story
Sun: energy of, 14, 36–37; as father, 29;
    relationship to, 25–26; workings of,
    13–15
surfaces, 33–34, 34f
sustainability, x–xi; versus economism,
    188–90
Swimme, Brian, 13, 26, 35
symbiosis, 68–71

Talbott, Steve, 84
Thich Nhat Hanh, 50, 107
Thys, Tierney, 103
ticks, 93
tigers, 230
time: relationship to, 25–26; and social
    capital, 217–18
Todd, John, 46–47
toilets, 46–47
Tolle, Eckhart, 174
Tomasello, Michael, 210
Transition Towns, 221–22
traumatization, 48
tree roots, 70–71
truthspeaking, 140–45
tubeworms, 101
Tucker, Mary Evelyn, 13
tundra, 125–26
Tzutujil, 210–11

umwelt, 58–59
universe: age of, 10; as alive, 21;
    Berry on, 5; end of, 12; history
    of, 18t; origins and workings of,
    6–12; reflection on, 26–27; and
    relationships, 20, 23–25
upcycling, 47, 234
Urquhart, F. A., 61–62

van der Ryn, Sim, 47
VanMatre, Steve, 52

Vilenkin, Alexander, 12
von Uexküll, Jakob, 59

Walljasper, Jay, 234–35
Wal-Mart, 193
wants, questioning, 169–70
waste, 45–47, 102, 104–5, 155, 224
water, bottled, 157–58
water capital, 130
water cycle, 41–43
Weingartner, Charles, 147
Werner, Patricia, 57

Wheatley, Margaret, 230
Wikipedia, 227
Wilber, Ken, 199
Wilcove, Dave, 93
wildfires, 127
wolf, 66
women, in Kerala, 151
worldview, 200–201. *See also* story

Zimmerman, Michael, 136
Zipcar, 222
zooplankton, 100

ABOUT THE AUTHOR

Christopher Uhl is a professor of biology at Pennsylvania State University. As a young man, he had an interest in both medicine and ecology. At Penn State he has been able to join these interests under the rubric of "ecological healing." During the 1980s, he studied the ways in which rainforest ecosystems heal following deforestation. Then, in the 1990s he focused on the role that universities might play in environmental healing by modeling sustainable ecological practices. He is also the author (along with his partner Dr. Dana L. Stuchul) of *Teaching as if Life Matters: The Promise of a New Education Culture*—a book reflecting a lifelong passion for teaching and learning. See Uhl's website at www.chrisuhl.net.